RACING

CHANGE

The End of

Product Innovation and Management

As We Know It

Shifting Our Focus to Managing

Product Lines as Complex Systems

PAUL J. O'CONNOR

i

Racing Change

The End of Product Innovation and Management as We Know It. Shifting Our Focus to Managing Product Lines as Complex Systems.

by Paul J. O'Connor

Published by

Advantage NPD, an imprint of

ADEPT GROUP Publishing, LLC,

Ponte Vedra Beach, FL 32082

www.advantagenpd.adept-plm.com

eBook ISBN: 978-0-9997381-2-2

Print ISBN: 978-0-9997381-3-9

U.S. trade bookstores and wholesalers: Please contact Adept Group Publishing, LLC, visit www.advantagenpd.adept-plm.com, or email the publisher at info at www.advantagenpd.adept-plm.com.

Printed in the United States of America

Publisher's Cataloging-in-Publication data: Business | Product Line Strategy | Product Line Innovation | Product Line Management | Artificial Intelligence | Technology Management

First Edition

Table of Contents

The knowledge, experience, tools, and thought worlds we've created for today's product innovation and management are mere prologue to an incredible epic that's just now begun.

- The Author

Forward

Racing Change

Why We Must Recast Product Innovation and Management and Shift Focus to Managing Product Lines as Complex Systems

Building Your Product Offering's Capacity To Perform and Transform in a World of Accelerating Change

Innovation is one of the most challenging and complex business issues of the 21st century.

Every CEO and executive team in medium-sized to Fortune 500 companies faces two conflicting challenges. They must deliver improved products more efficiently while also commercializing new offerings that align with emerging market and technological opportunities. Never have large companies encountered such intense, simultaneous challenges. These firms, many of which lead their respective industries, must both perform and transform.

The situation arises because our methods have changed little while everything around us is evolving at an increasingly rapid pace. Not only is change accelerating, but the rate of acceleration itself is also increasing, and it appears that no end is in sight.

We've already entered the logarithmic upswing of "adjacent possible," as described by Stuart Kauffman.[1] For products, adjacent possible technologies and business elements can combine to create new technologies and other business components, which then merge to form even more adjacent possibles. The upswing suggests a chain reaction in creating new things that are within our ability to achieve or deliver.

Change Acceleration

Experts also declare that Moore's Law—doubling CPU chip performance in eighteen months—is too slow.[2] Designing more components for each chip has become irrelevant due to AI's massive parallel scaling of GPUs. Moore's model, which we once believed was a significant driver of change, now seems almost prehistoric.

AI's rapid advancement and adaptation over the past few years have transformed from simple large language models into more sophisticated compound AI systems that integrate facilitated prompts, reasoning, planning, and action-taking.[3,4] Producers of large language models have already claimed to have developed artificial general intelligence (AGI), capable of automating much of people's work. Applications are expected to become commonplace by the end of 2026.[5]

Artificial Superintelligence (ASI) models, representing a significantly higher level of intelligence, will resemble a lecture hall filled with super-geniuses. These models will perform tasks we never thought possible, leaving mere mortals struggling to understand their analysis, choices, and actions.

At the current staggering pace of advancements, AI experts predict a breakthrough in ASI within the next two to three years when AI recursively self- improves, writing its own code and implementing its own procedures to enhance its performance. Humans simply step aside and let it run 24/7. I know; that's scary.

The impact of ASI will be astounding as it generates everything from biological molecules and chemical processes to novel materials, complex software, specialized hardware, and optimized CAD/CAM designs. We may witness the initial stages of ASI's expansion into large-scale corporate research and development before 2028. This immense knowledge enablement will fundamentally transform product innovation and management.

The AI Change Tsunami

AI will trigger a wave of change alongside remarkable advancements in robotics, automation, sensors, energy generation, and soon, quantum computing. This transformation will lead to exponential increases in productivity and intelligence, disrupting products and innovations everywhere. Companies that cannot keep pace will not survive.

While many people are raising urgent concerns about the ethics, trust, and societal well-being associated with AI, particularly ASI, this book serves a different purpose. I aim to clearly articulate how these changes, especially their accelerating pace, influence product innovation and management in medium to large-sized companies.

AI advancements will not only impact your products and organization; they will also affect your partners, competitors, customers, and all participants in the global economy. Everyone is feeling the influence of AI-induced opportunities and challenges. Unfortunately, management teams that rely on today's product innovation and management methods to navigate the tidal wave of change will not survive, even if they leverage AI to enhance their outdated techniques.

The Efficiency-Change Dilemma

The main issue is that the methods large companies use for product innovation and management impede their ability to keep pace with rapid changes. Management teams shouldn't expect to "best practice" their way through this immense task.

The race is not only against competitors; it's also a race against change itself!

Few companies possess the knowledge and skills to tackle accelerating change. While most companies excel at incremental developments, few succeed in achieving major innovations. What's most alarming, however, is that the speed of change greatly amplifies the difficulty of this challenge. I refer to today's situation and the inherent challenge as the **Accelerating Change Race.**

The race is not a dilemma of competing demands pulling a company in different directions. It's not a choice between efficiency and change. Instead, companies must achieve both.

Many boardrooms are grappling with this issue. Shareholders are becoming increasingly vocal, urging company leaders to enhance performance and transform their organizations and products. These tasks are substantial yet essential.

The Big Split

Today's common approach to the challenge is organizational, dividing efforts into two paths: one to support current product management and another to foster major innovations.

The first course in this traditional approach emphasizes best practices that enhance existing offerings. The objective is to lower costs for current products and execute incremental innovation projects more swiftly and with less waste. Experts refer to this as sustaining the core.

The second course involves establishing entities that operate independently from the core business and current product offerings. Because their management reports solely to the corporation's top leaders, experts refer to this work as "Corporate Innovation." The mission is to "create the new." It addresses unmet market needs and delivers new technologies and innovations without distracting the core business from its focus on efficiency. This concept has been in place for several decades.

A House Divided

We're told that the two approaches, whether with or without AI enablement, are incompatible, and companies must keep them separate. However, the traditional two-prong approach is costly, slow, and ineffective. Most product innovation and management experts assert that the best course of action is to reinforce or double down on the split approach. Sadly, this direction is misguided.

The larger issue is that the split approach fails to generate significant innovations for existing product lines. The results do not help current lines address the massive changes that AI will bring or the drastic shifts in customer needs intensified by instant communication and knowledge sharing.

Unfortunately, today's divided approach stifles the possibilities and opportunities necessary to transform current product lines.

Poor Performance

While today's Corporate Innovation can assist companies in developing new businesses, few of these ventures have been substantial or successful enough to allow companies to transition *away from existing products. Instead, "creating the new" has yielded notably poor results, compelling companies to seek greater gains from their current offering.*

The only positive news is that competitors likely share the same poor track record. The few large companies achieving successful results in Corporate Innovation are well-positioned to serve rapidly expanding markets. Tesla, Apple, and Amazon lead the list. However, it is not a coincidence that these firms also drive significant changes in their core products. Their success is largely due to their ability to understand technologies and markets and to formulate smart strategic moves. It does not stem from separating major innovations from core offerings.

Separation is irrelevant if teams cannot identify strategic moves that drive greater change in core products and improve the effectiveness of Corporate Innovation. This role involves more than simply generating ideas, innovating, and scaling a product.

Dispel Faulty Assumptions

Unfortunately, many leadership teams fail to recognize that their divided approach is inadequate. Instead, they embrace the idea, championed by numerous Corporate Innovation advocates, that existing product lines cannot change and evolve.

This notion is incorrect, and we must dispel it. The situation is not unchangeable; however, large companies need to update their outdated approach to product innovation and management.

The Solution

In this book, I explore the reasons and methods for helping teams and organizations overcome the challenge of pursuing efficiency and change simultaneously. The approach requires a deep understanding of a product line's composition and behavior. To succeed, we must transform the very context in which we develop and manage products, ensuring that innovation and efficiency work together.

This significant shift in perspective encourages those aiming to create "the new" to reconsider their strong urge to develop a business model and quickly scale processes immediately after evaluating a concept. The better approach is to focus on and invest in building product line systems with a clearly defined structure or, as I share, carefully designed architecture. Building a multi-product system is crucial for scaling a successful business.

A Complex System

Product lines are complex systems. This complexity renders traditional product innovation and management methods ineffective when teams aim to enhance a product line's performance while evolving, advancing, and transforming their offerings to stay competitive.

Concurrent efficiency and change necessitate a reinvention of product innovation and management. We must move away from traditional single-product thinking, which emphasizes developing one product to satisfy a specific set of customer needs. This crucial insight lays the groundwork for effectively building, advancing, and managing product lines.

Our Thinking and Action

This book shares the key to achieving greater efficiency and facilitating significant change. It focuses on understanding and managing product lines as complex systems rather than merely concentrating on single-product orderly systems in which cause-and-effect relationships dictate outcomes. Such understanding isn't about the organizational, leadership, or attitudinal mindset changes that many experts advocate. Instead, it offers a solution grounded in knowledge, informed decision-making, and the execution of strategic actions.

Throughout the book, I examine topics related to people. However, I do so to illustrate their critical roles as enablers of the product line system, not as solutions. Because change is now continuously accelerating, even the best leaders and organizations will not escape the failures that will arise if their teams and managers do not adjust their thinking and actions.

Managing a product line as a complex system changes how teams view their products and services. A single-product perspective transforms into a dynamic multi-product landscape. This shift creates opportunities to leverage cross-product relationships and design a coherent series of strategic moves, which are essential for delivering performance and adapting to rapid changes.

A Context Shift

The concepts and principles I share in this book come together to create a powerful framework. This represents a significant shift in context, which I refer to as Product Line Systems Management (PLSM). It serves as a guide for navigating and leveraging even the most uncertain and dynamic situations. It also encourages learning from and building upon a company's existing methods and practices.

The PLSM framework guides teams to experiment with and creatively explore the interaction of a product line system's components and agents. Doing so drives innovation. Importantly, the framework promotes enhancing the line's architecture while closely considering the system's limitations. These aspects are essential for achieving superior performance and realizing significant advancements that please customers and motivate competitors to seek new avenues.

PLSM succeeds because it is grounded in sound practices and robust theory. Its foundation is solid, not just a lot of consulting nonsense.

Adapting PLSM

Product Line Systems Management sets itself apart from what I, along with other advisors, have learned, and from the teachings of academics over the past several decades. It presents models and principles that differ from those many of us have relied upon throughout our careers. I fully expect that most consultants, software vendors, and academics who depend on single-product, orderly systems thinking will resist or overlook the necessary change. That reaction is natural.

Detaching your thinking from the paradigms and heuristics you have spent years learning and adopting is a challenging task.

Most importantly, PLSM enhances efficiency and increases the capacity for change in existing products and services. It directs the core business to improve performance and support product and organizational transformation, rather than solely focusing on corporate innovation and "leaning out" existing offerings. This book aims to explain and demonstrate why and how to do just that.

You'll also read how several so-called "best practices" embedded in common product innovation and management approaches can hinder a product line's performance and ability to adapt. The framework helps teams identify and modify these practices. In each of the instances I share, I also explain why such best practices may be far from the best today and in the future.

About Me

I feel fortunate to have enjoyed a rewarding career that has focused on various aspects of product innovation and management. I've collaborated with product managers and teams, and I've mentored senior executives in some of the largest companies in the world.

Throughout my career, I have had the opportunity to learn from many top academics and experts in the field, incorporating their theories and methods into my thinking. It has been a privilege to know and learn from many of the leading figures mentioned in this book. I have contributed to developing and refining processes at major companies, working with teams to drive innovation, establishing processes, manage portfolios, set up roadmapping and support IT systems, develop strategies, and address challenges related to product lines. You can learn more about me at the end of this book.

About This Book

This book challenges many of the product innovation and management beliefs I've held sacred. I encourage anyone who governs, advises, leads, or contributes to products and innovation in medium and large-sized companies to read it and reflect on what I've presented. In the book, numerous new ideas and strategies are presented that should interest product managers, marketers, product developers, engineers, technologists, and strategists. As you read, you'll notice that I address topics to product line managers, their teams, and business unit leaders.

I simplify the content by discussing change and PLSM, focusing on existing product lines. Additionally, I address the crucial role of product line systems in Corporate Innovation when they support my argument. Chapters 27 and 28 illustrate how PLSM serves as a powerful, yet notably overlooked, facilitator of Corporate Innovation. It provides essential guidance for initiatives aiming at "radical" or "breakthrough" innovations.

Understanding a product line system's architecture, components, forces, actors, constraints, enablers, and constructors can be challenging. This is especially true when someone is unfamiliar with the company or its product lines. Therefore, the book features narratives about well-known companies and their product lines. I have moved cases that require deeper exploration to the Appendix to avoid overwhelming readers.

Appendices, Narratives, and Glossary

Please note that an appendix detailing several topics and case study narratives is included at the end of this book, as well as another online at https://adept- plm.com. The same applies to the glossary; one is located at the end of the book, while the other is accessible online. Additionally, the online content includes templates and "get-started" guides to assist with the transition to PLSM.

Our Journey

Welcome to an exciting journey into Product Line Systems, their creation, and management. I hope you find this book insightful and helpful as we navigate our new world of rapidly accelerating change.

- Paul O'Connor

Acknowledgments & Thanks

I learned long ago that individuals working in product innovation and management are unique. I've spent my entire career in this field. It began in 1982 when I attended a meeting of the New York City chapter of the Product Development and Management Association (PDMA). At that time, PDMA was only a few years old and heavily focused on marketing academics whose thinking was primarily directed at large companies.

Those members who were service providers, like me, sold "innovation-in-a-box," creativity workshops, or team-building services. Few of the products we discussed included software components, and no company offered software to support product development or management. There was no concept of Stage-Gate, Agile, Design Thinking, Portfolio Management, Open Innovation, or Roadmapping. MS Project didn't even launch until a decade and a half later.

At PDMA meetings, speakers conveyed one of two messages: be innovative and take risks or learn to foster cross-functional teamwork. Overhead transparencies were common, and every presentation featured images of boxes outside of which we were told to think, as well as factory chimneys and silos we were told to tear down. Looking back, though, the 1980s and 1990s were productive years for the community.

I was privileged to have a front-row seat to the product innovation playing field by serving on PDMA's board and becoming its president. This position introduced me to many intelligent individuals who influenced and shaped product innovation into what it is today. From these early contributors in the field, I learned what seemed to work and what didn't.

I also observed many approaches, practices, and tools that failed to take hold, despite my initial confidence in their success. You can add QFD, NewProd screening, intrapreneurship, and Theory of Constraints to that list. I also watched things take hold, even when I was certain they were misguided. Long ago, I came to realize that something was occurring in product innovation and management that I, along with many others, clearly did not understand.

My understanding of product innovation and management comes from academic research, direct experience, and the many narratives shared openly by others. I am grateful to my friends and colleagues at PDMA for this incredible learning experience.

My deepest gratitude goes to the practitioners who questioned and sometimes rejected what I was saying, teaching, selling, or helping companies to implement. Their refusal to accept what many leaders in the community viewed as a rock-solid product innovation and management approach has been a crucial force in advancing the field.

Thank you to Will Smith at Black & Decker for pointing out that the Stage-Gate process is single-product oriented when, in reality, he and others in large companies must manage a changing portfolio of projects. We clearly understand Will's perspective today, but at the time, it raised some eyebrows.

That was in 1992, and it led me to focus on portfolio management, which, in turn, led to roadmapping and an understanding of product lines. This experience marked the beginning of multi-product thinking; a key topic extensively explored in this book.

Today, economically driven "lead users" (thanks to Eric Von Hippel) are actively advocating for radical innovation (thanks to Gina O'Connor and Mark Rice). Unfortunately, despite everyone's efforts, this path has not been as successful as most players had hoped.

At the risk of facing boos from the innovation community, I must say we've fallen significantly short in executing major innovations. The reason, we're told, is that the uncertainty and long-term nature of such innovation deter top management. I've heard that for over four decades.

To this day, I still find it surprising how companies like Kodak and Polaroid failed. I had the privilege of engaging with these firms and experiencing the cultures that radiated innovation. These companies attracted the best and brightest college graduates. Yet, their histories show that even the combination of highly intelligent people and supportive cultures was not enough. Something much greater was needed then, just as it is today.

I wouldn't have survived without some success over my four-plus decades in product innovation, development, and management. I've worked with numerous clients worldwide, providing guidance to enhance their product innovation and management practices. I am grateful to each of these companies for the experience they offered me.

One person I do not know personally deserves recognition as a significant influence for this book. Dave Snowden of The Cynefin (pronounced kuh-NEV-in) Company is an expert and an outstanding teacher in the field of complexity science and complex systems. The depth of my exploration into the issues we face in product innovation and management would have been considerably limited without Dave Snowden's prolific and thought-provoking insights into complexity science. Thank you, Dave.

I extend my deepest gratitude to my wife, Donna, for her unwavering support. Her extensive experience—from teaching creative writing to college freshmen, to coaching leadership at McKinsey and Company, and tackling organizational development at various Fortune 500 companies—has made her a trusted partner in discussing the issues I explore throughout this book. She is also the love of my life. Thank you, Donna.

"The pace of change has never been this fast, yet it will never be this slow again."

- Justin Trudeau, Canadian Prime Minister, Davos Keynote Speech

"All you need to know to understand which company will win a technology competition is each company's first and second derivatives of the rate of innovation."

- Elon Musk, one of many Twitter proclamations.

SECTION ONE

The Innovation Problem

"Ask not what's inside your head, but what your head's inside of."
- William Mace, Professor of Psychology, Trinity College

Context is everything. We must rethink how we approach innovation and product management. Our old methods, tools, and processes don't stand a chance. - *The Author*

Chapter 1

The Accelerating Change Race

The Problem We've Created By

**Avoiding Incremental Failures and
Splitting off Major Innovation**

Al executive teams have concerns about their company's products. They want the best for the least money and don't want to fall behind competitors. In response, we've pushed as hard as possible on "best practices." These are the methods that experts advise will help companies outperform their competitors. Unfortunately, this approach hasn't produced the results we had hoped for, and our prospects may be worse.

Our problem arises from the way the product innovation and management community defines the issue. We've declared it to be two separate challenges. One involves driving greater efficiency into core product offerings, while the other focuses on augmenting the business with major new offerings that leverage new technologies to meet unique customer needs. We distinguish between sustaining the core and creating the new. See Figure 1.1.

Experts rarely recommend that companies alter their core offerings. Instead, they suggest making small adjustments that enhance these products while keeping significant innovations separate from the core. This approach has been in place since before most of today's managers began their careers.

The common practice separates a focus on efficiency from a focus on innovation. However, the world has changed significantly since this practice became prevalent, and that change is set to accelerate rapidly as artificial intelligence proliferates and evolves. The trajectory resembles a cliff more than an upward slope. Such rapid change makes common product innovation and management strategically misguided and economically harmful.

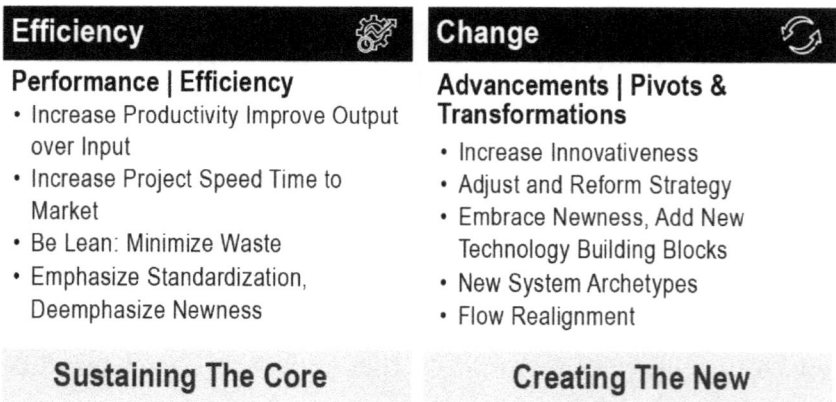

Efficiency	Change
Performance | Efficiency	**Advancements | Pivots & Transformations**
• Increase Productivity Improve Output over Input • Increase Project Speed Time to Market • Be Lean: Minimize Waste • Emphasize Standardization, Deemphasize Newness	• Increase Innovativeness • Adjust and Reform Strategy • Embrace Newness, Add New Technology Building Blocks • New System Archetypes • Flow Realignment
Sustaining The Core	**Creating The New**

Figure 1.1 Organizations pursue two conflicting orientations: efficiency and change. The problem is that when a company focuses its methods on achieving one, it undercuts its ability to carry out the other.

Artificial Intelligence Changes Everything

AI is a remarkable technology. However, its stunning rate of advancement is creating challenges for businesses, product lines, and individuals alike. Most concerning is the fact that this pace of change is accelerating with the emergence of Retrieval Augmented Generation and Automated Reasoning. These AI enhancements indicate that we're on the verge of company-wide AI agents capable of performing many tasks traditionally handled by humans.[6] This represents business-specific General Artificial Intelligence. Before 2030, we will witness even more powerful business-specific AI guiding us toward possibilities we've never envisioned before.

See Appendix 1 for an in-depth discussion on the incredibly rapid advancements of AI.

The speed and magnitude of change make the old-school approach highly ineffective. It creates a new challenge that traditional methods and practices cannot address. I call it the **Accelerating Change Race**. We can no longer afford to prioritize efficiency over change for our core products.

To be blunt, AI will make the old, bifurcated thinking about product innovation and management obsolete. The challenge is not a dilemma pitting efficiency against change; it's about finding a way to keep pace with an ever-increasing rate of change while consistently delivering efficiency.

Many Views, Same Recourse

The concern was, and still is, that static products become outdated. It's unavoidable. Eventually, companies face severe consequences if they can't change their core products. Most managers recognize this but believe they have limited options.

Over the last several decades, experts have tagged their view of the efficiency-change challenge differently. See Table 1.1.

Table 1.1 Recognition of The Old School Efficiency-Change Dilemma

Issue Statement	Discussion
Ambidextrous Organization	A well-accepted organizational approach toward solving the efficiency-change quagmire. First shared in HBR 2004 by Stanford Professor Charles O'Reilly and Harvard Professor Michael Tushman.[7]
Innovator's Dilemma	Harvard Professor Clayton Christensen shared his view on the issue in his 1997 book entitled "The Innovator's Dilemma."[8]
Lean Startup	In 2011, Eric Ries developed an approach that combines entrepreneurship and venture capital and applied it to large corporations.[9]
Intrapreneurship	Gifford Pinchot raised the issue and proposed his solution in his 1985 book Intrapreneuring: Why You Don't Have to Leave the Corporation to Become an Entrepreneur.[10]
Radical vs. Incremental Innovation	First declared in 2000 by RPI professors Gina O'Connor, Mark Rice, and their colleagues, this seminal work is foundational to corporate innovation.[11]
Performance vs Dynamic Capabilities	This view highlights a capability gap between core offerings that aim for efficiency and corporate innovation that seeks to leverage significantly new opportunities for the company.
Dual Innovation	This approach is similar to and supportive of the Ambidextrous Organization but adds a middle ground to foster change to the core offering.[12]
Corporate Innovation	The organizational approach for large organizations seeking to separate radical or disruptive innovation from the core business and core product offerings.
H1, H2, and H3 Horizons	McKinsey and Company introduced this model in 2009. It states that a core business should only tackle near-term ready technologies (H1) and push longer-term ready technologies (H2 and H3) to Corporate Innovation.[13]

Brilliant Approaches but Elusive Results

It's hard to argue against any approach proposed by the best and brightest minds. These experts share a common goal and base their guidance on what appears to be intelligent analysis. Each one strives to overcome the dilemma. Sadly, none are sufficiently effective. More concerning, the new age of accelerating change renders them only minimally relevant. Still, the proponents of each continue to blame the ineffectiveness on poor management buy-in or an undisciplined organization.

Stanford Professor Steve Blank critiques the unsuccessful attempts on the change side of the dilemma, labeling it Innovation Theater.[14] This term effectively illustrates his point. Everything appears impressive and seems positive, but business results remain short-lived. It's not due to a lack of effort; rather, what companies are doing isn't yielding the results that leadership teams and shareholders desire.

The late Harvard Professor Clayton Christensen presented us with eight principles he referred to as the Innovator's Solution.[15] But his principles are not flawless. One tenet is to prioritize improving cash flow initially. Christensen explains that positive cash flow enables an innovation initiative to withstand organizational challenges. He also recommends starting small rather than trying to dominate an industry.

Neither of Christensen's principles yields meaningful results against the magnitude of change we face. These rules did not guide Amazon or Tesla. They don't assist businesses that have access to low-cost capital and large investments. By default, Christensen's principles exclude innovations that require significant scale to succeed.

More Than a People Issue

As far back as 1985, Gifford Pinchot introduced the concept of intrapreneurship to help organizations overcome innovation challenges. Pinchot encouraged practitioners to break the rules and norms of business. To innovate, he preached, you need a "just-do-it" attitude because, as the saying goes, "it's easier to ask for forgiveness than permission." (On a personal note, Gifford and I worked at a small innovation service firm just before he published the book that launched his radical approach to innovation. I've always viewed intrapreneurship as a response to our struggle to get clients to develop the concepts we created. I worked within the intrapreneurship community for a short time before teaming up with Professor Robert Cooper from McMaster University. Professor Cooper was masterful at driving efficiency in core product development.)

Many companies embraced intrapreneurship. Unfortunately, it resulted in activities that were disconnected from the norms of large corporations.

In large companies, major innovations require size and scale to succeed. Do you think an intrapreneur at Ford could have successfully brought a profitable electric vehicle to market? Even Ford's dedicated organization still struggles. What benefit would an intrapreneur provide? Unsurprisingly, Babson College Professor Gina O'Connor, a leading expert in radical innovation and the ambidextrous approach to Corporate Innovation, declared the power of Intrapreneurship to be a fallacy.[16]

The Ambidextrous Organization

O'Reilly and Tushman adopted a different approach. They advised companies to become "ambidextrous" to tackle the two-pronged dilemma. However, this has also proven insufficient, particularly when competing with skilled adversaries. We require much more than today's two-pronged strategy.

Separating major innovation efforts from the core business may foster the development of new products, but this independence rarely scales effectively. Sustainable, bespoke businesses seldom emerge. Given the disappointing results over several decades, I feel confident in declaring that the ambidextrous approach is sadly lacking and inadequate for keeping pace in the Accelerating Change Race.

Corporate Innovation's Shortfall

Frank Mattes, a European consultant specializing in the split approach, cites research indicating that corporate innovation success rates are less than three percent.[17] Mattes calculates this minuscule rate by combining studies from Bain and Company and Accenture, which measure success as building a continuous business that generates at least $50 million in annual revenue. A quick cost analysis shows that large corporations must quadruple their success rate if they want their corporate innovation efforts to be economically viable. This presents a significant challenge for a task that big companies have struggled with unsuccessfully for over five decades. Incremental improvements from adopting the Corporate Innovation "best practices" that Mattes recommends may help, but they are unlikely to quadruple the success rate. Something much more significant must occur.

Corporate innovation success must also involve transferring innovations to existing core product offerings. However, such transfers in today's divided approach are often problematic because the "efficient" core business to which they must be transferred has little capacity to change. Yet those working in

Corporate Innovation do not see it as their responsibility to enhance the Change Capacity of a core product line. (Appendix 5 discusses a product line's Change-Capacity. Understanding it requires a basic knowledge of product lines as systems, which is shared throughout this book.)

Lean Startup's Misstep

When Eric Ries introduced the Lean Startup approach over a decade ago, it energized managers in large organizations.[18] Many embraced the idea of being entrepreneurs in corporate settings. Unfortunately, like intrapreneurship, the Lean Startup movement does not succeed in large companies. Just ask Jeff Immelt, former CEO of GE. He invested heavily in the approach, even bringing Eric Ries into his company to guide his multibillion-dollar investment in it.

Immelt's bet failed miserably, forcing GE shareholders to oust Immelt. Ries was left to declare that the Lean Startup approach doesn't fit large corporations. The ill-fated movement was also a major factor prompting Stanford Professor Steve Blank to call Corporate Innovation "theater."

In his 2019 Harvard Business Review article, Professor Blank also clashed with the formidable McKinsey & Company by pointing out their flawed logic.[19] The consulting firm encouraged companies to categorize new technologies into H1, H2, and H3 time horizons. Blank clearly states that this approach is a significant mistake. The McKinsey model advises managers to shift mid-term (H2) and long-term (H3) technologies and innovations away from the core business, leaving only incremental (H1) technologies to drive the core business.

Professor Blank argues that there is an increasing necessity to integrate H2 and H3 technologies into core business operations more rapidly than the McKinsey model suggests. He asserts that the self-imposed time frame isn't what matters. Instead, it is the demands of customers and competitive dynamics that necessitate earlier adoption. Relying on Corporate Innovation initiatives to establish fully developed businesses using H2 and H3 technologies is unwise, especially as change accelerates.

Embracing McKinsey's three-time horizon approach complicates the efficiency-change dilemma. If core products do not advance sufficiently, they will decline even more quickly. It is a mistake to restrict core product development to only H1 technologies. H2 and H3 technologies must also impact and drive core products forward. However, to accomplish this, the core business must transform its failure-averse strategies and adopt new practices.

The Negative Innovation Loop

Every large company feels pressured to change and transform its products, not just to adapt and modernize them. This pressure intensifies as everything accelerates. Customer needs are evolving more rapidly, and competitors are developing new offerings at an increasingly fast pace. Moreover, we observe that technologies such as AI, ML, and data analytics are rapidly being integrated into practice. However, alleviating this pressure isn't simple, especially given the divide between Core Products Efficiency and Corporate-Breakthrough Innovation that many practitioners adopt.

Companies that double down on core product efficiency create an organizational force that resists change. Many experts, including myself, refer to this force as "efficiency inertia." This powerful force prevents core businesses from pursuing significant innovations. Unfortunately, it also generates an even greater need for impactful Corporate Innovation.

Because Corporate Innovation intentionally maintains independence from, rather than alignment with, core products, its outcomes are unlikely to match or enhance current offerings or arrive soon enough as customized product lines to remain relevant. This gap compels leadership teams to increase demands on their core products and redirect resources away from Corporate Innovation. It creates an ongoing cycle I refer to as the Negative Innovation Loop. See Figure 1.2.

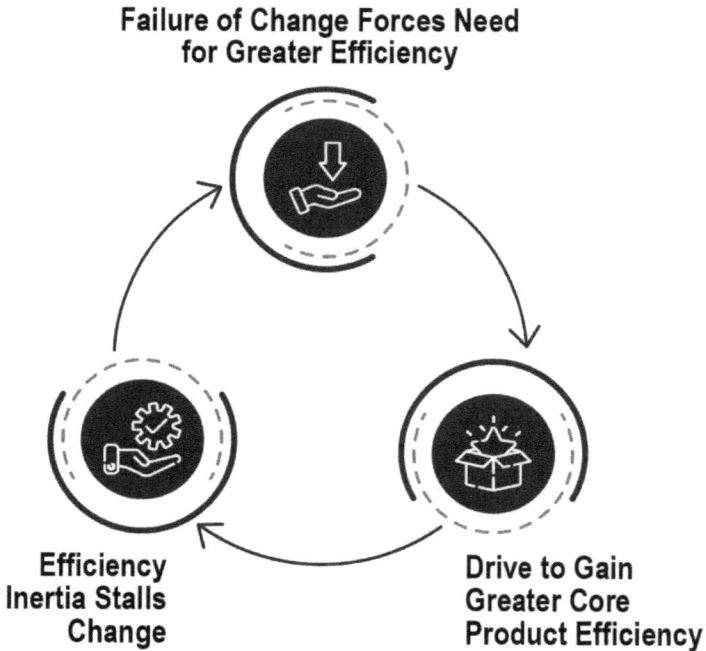

Failure of Change Forces Need for Greater Efficiency

Efficiency Inertia Stalls Change

Drive to Gain Greater Core Product Efficiency

Figure 1.2. The Negative Innovation Loop *is a systemic feedback loop in which efforts to increase efficiency in core products reduce the organization's ability to innovate, which pressures core products to support innovation. This loop perpetuates the underperformance of corporate innovation and the core product's march into maturity and decline.*

Some companies weather the negative cycle because their current offerings possess stickiness or brand appeal. Unfortunately, product stickiness tends to fade as life cycles mature, often resulting in stagnant and declining sales. Also, brand appeal tends to be fleeting when product performance and perceived value fall behind those of competitive alternatives.

Proponents of Corporate Innovation exacerbate the problem when they claim organizations must improve Corporate Innovation because their core business and its core offerings cannot change. These well-meaning experts categorize the core business as fixed. They suggest that those managing the business's bread-and-butter products and services should drive efficiency and generate greater cash flow to support their innovation efforts.

The problem lies in the assumption that a core business and its offerings cannot change. This is a false belief. Core businesses can evolve and change, and we observe this frequently. The real issue is that, over the years, our traditional management approach has diminished the capacity of core offerings to adapt. Unbeknownst to many of us, the Corporate Innovation expert's assertion that core businesses cannot change has contributed to this erosion.

The effects are severe when companies accept the Negative Innovation Loop as normal and without recourse. Behaviors and judgments reinforce poor outcomes. Collectively, these outcomes make it increasingly difficult to break the cycle at a time when doing so has become critically important.

The Negative Innovation Loop is an unintended consequence of emphasizing best practices to avoid failure in the core business while diverting significant innovations from core product offerings. It's a classic negative feedback loop, an effect that many innovation experts never anticipated.

Change In the Trenches

Consider the impact of AI on internet search, protein synthesis, adaptive automation, and FinTech.

Google must transition its core search tools to be AI-enabled. A search technology that delivers search summaries without advertisements poses a threat to the company's pay-per-click business. This development significantly reduces the importance of an efficiency push within the core business.

Google, Alphabet Inc.'s largest business unit, faces a dilemma as Microsoft Bing-ChatGPT challenges Google Search directly, while companies like Perplexity and You.com circumvent pay-per-click advertising altogether. A one percent market share equates to approximately $1.7 billion in revenue, which translates to about $7 to $10 billion in equity valuation. In 2020, Google commanded a ninety percent market share. Do the math. A collapse or significant reorientation of the traditional search market could have disastrous consequences for the company and its shareholders.

Most experts agree that Google would be unwise to allow its core search tools to decline while establishing a separate business to compete with AI-powered search tools. If they were to take this approach, they would enable competitors to capture significant portions of the search market, ultimately falling too far behind in the AI-enhanced search landscape. This strategy could jeopardize the stability of Google's core business.

Then there is Google's DeepMind company. With its AI-driven advancements, the company has disrupted multiple industries.

This advancement impacts every drug company, clinical testing firm, and medical practice involved in large-molecule protein discovery, testing, and synthesis.

Recently, Google DeepMind and Isomorphic Labs announced AlphaFold 3, an AI model that accurately predicts the structure of proteins, DNA, RNA, and fatty lipids. This model is so significant that DeepMind's co-founders, Demis Hassabis and Dr. John Jumper, were awarded the Nobel Prize in Chemistry.[20] It's a major advancement for many stakeholders in drug development, biology, and medicine.

Big auto companies face similar challenges. To compete with Tesla and Chinese firms like BYD, Geely, Nio, and Xpeng in the electric vehicle market, car manufacturers must adopt adaptable automation, new battery technologies, and other advanced technologies. They must change.

The challenge seems overwhelming. Ford, GM, and Toyota express their frustrations about competing with Tesla and BYD in the EV market. BMW, Mercedes, and VW share similar concerns. Many industry experts even assert that the German auto industry is on the brink of collapse.[21]

The issue isn't about delivering a groundbreaking electric vehicle that impresses customers; it's not an inability to penetrate the EV market. Instead, the problem is that when they believe they can significantly impact the market, the leaders like Tesla and BYD adapt even more rapidly with innovations such as adaptive automation, gigacasting, and solid-state batteries.

The dynamics are significantly more intense and consequential than those faced by previous leadership teams in the auto industry. I'm sure even the brightest leaders in these companies have questioned their thinking and actions.

We also see banks of all sizes facing the same challenge. They must reshape their core operations to include instant cash transfers, mobile banking, digital currency, predictive modeling, seamlessly integrated accounts, and numerous customer-centric capabilities. At the same time, cybersecurity concerns continue to grow each year. What will happen when cybercriminals fully adopt AI? Will they exploit Artificial Super Intelligence, a capability many times smarter than even the brightest human?

In each case, major changes must be made to core product offerings, rather than being separated into Corporate Innovation efforts. This shift is what Professor Blank argued. Companies must integrate H2 and H3 technologies into their core product offerings more quickly and effectively.

Failing to Change

Things change, and today that change is accelerating. No matter how effectively product managers and contributors deliver greater efficiencies, customers always move on to better solutions. You face serious issues if you don't offer the solutions customers want. If left unattended, failing to adapt your product offerings will send your company into a tailspin.

In one industry after another, the Negative Innovation Loop is exacerbating the efficiency-change tug of war. Large companies must prevent the negative loop from further undermining their market incumbency.

Change Both the Core and Corporate Innovation

Keeping pace with the Accelerating Change Race requires us to rethink our dual approach and revise our methods. In the following chapters, I will explain why and how we need to shift our thinking from a single-product, orderly systems perspective to a multi-product, complex systems perspective. This transition places product lines and their management at the center. You will find that this change significantly impacts the race against accelerating change.

Product lines are systems we need to influence to break the negative innovation cycle. They present the greatest opportunity to progress with and alongside the rapid changes every company encounters.

My perspective and practices differ from the standard approaches that companies employ. But that's the point. Pushing harder on the same methods, even copying and using "best practices," won't achieve the results that big companies need.

In the next chapter, we will explore why product lines are crucial and examine how our approach to products and product lines influences our ability to keep pace with the Accelerating Change Race.

The Accelerating Change Race

The accelerating change of technologies, customer needs, and industry forces has impacted how companies develop and manage products. Companies must improve and adapt their core products while creating entirely new offerings that meet customer needs currently unmet by their business. Splitting the response into two separate efforts—one for efficiently delivering core products and the other for generating substantial growth—is inefficient and ineffective.

The response to the Accelerating Change Race should not be divided between efficiency and change. Instead, it should take a more impactful approach. First, companies must enable change in their existing product lines. Second, companies need to enhance their ability to create and scale product lines, rather than concentrating on single-product innovations and then shifting to building businesses.

The old-school split approach is poorly suited to a fast-changing world. Companies must develop the ability and capacity to adjust core products while revising their approach to radical innovation. This requires a transformation from outdated single-product thinking to more dynamic, multi-product thinking.

Chapter 1 Summary

The Accelerating Change Race

- **Efficiency-Change Dilemma:** Big and medium-sized companies must make their current products perform superbly and do it efficiently. But now, accelerating change has redefined this job. The old approach pushes in opposite directions. However, core products must change, and radical innovations must focus on supporting existing product lines and creating new ones. Unfortunately, most companies are ill-equipped to take on this approach. This problem, caused by accelerating change, forces us to rethink how we develop and manage products.

- **Old Ways of Doing Things:** In the past, core business functions focused on improving current products and left major innovations for an independent part of the company. We've used Lean, Agile, Design Thinking, and Portfolio Management practices to improve core products. But these techniques don't help enough to enable companies to create bespoke new businesses proficiently.

- **A Negative Innovation Cycle:** Launching major innovations is tough. Doing it in a corporate setting is even tougher; most companies cannot do it. The lack of success causes greater pressure to develop and extend core offerings more efficiently. In turn, the situation increases "efficiency inertia," which works against developing major innovations. This interplay between core products and major new offerings creates a negative cycle that works against innovations.

- **Core Products Matter:** Small improvements to core products aren't enough. Companies must change their core products markedly to keep up with competition and technological advancements.

- **Fixing the Problem:** Big companies must build the capacity and capability to change core products to keep pace with the Accelerating Change Race. Many managers' big departure is that they must embrace changing existing products just as they embraced driving their efficiency. For decades, though, such change had been seen as implausible.

Chapter 2

Why Product Lines?

Indivisibility, Granularity, Practicality

We need to address two major questions as we confront rapidly accelerating change:

1. *Why must we reinvent product innovation and management? and,*

2. *Why should we focus on product lines?*

Recasting Product Innovation and Management

Several overlapping factors render our current product innovation and management practices ill-suited for succeeding in the accelerating change race. This is especially true when using AI to enhance current practices, as it exacerbates the critical problems inherent in the old methods.

I have already discussed how current practices avoid major changes to existing offerings and keep such efforts separate from the core business by creating independent entities. This philosophy doesn't bode well for existing lines, the "bread and butter" of big companies. Surviving in the rapidly changing landscape is less likely if core offerings remain unchanged.

When computers entered the workforce in the 1980s and 1990s, everyone expected a surge in productivity. However, that surge did not materialize until companies embraced organizational and digital transformations. Still, despite this, a great deal of unproductive work persisted throughout the process.

Anthropologist David Graeber highlighted this in his much-debated book "Bullshit Jobs, A Theory."[22] BS jobs, Graeber argues, are those that do not contribute to society or individual well-being. He asserts that these jobs emerged due to "managerial feudalism."

Such kingdom-building was an unintended consequence of the first wave of computers that Graeber says enabled the financialization of every aspect of the economy. Feudalism, he states, is driven by the pursuit of money and power, not societal or personal well-being. To quote Graeber, financialization created "dumb jobs," such as "HR consultants, communication coordinators, PR researchers, financial strategists, or the sort of people who spend their time staffing committees that discuss the problem of unnecessary committees."

Empirical research on the topic indicated that BS work was real, with approximately 20% of respondents stating that their jobs were socially useless.[23]

The AI-induced Accelerating Change Race creates a different problem. It isn't BS jobs; the problem is **"BS" products**.

Such "enshittification" was highlighted by British blogger Cory Doctorow in his November 2022 post, which was republished in the January 2023 edition of Wired. It was entitled "The

Enshittification of TikTok."[24] For online products and services, Doctorow declares that Product Enshittification stems from "the ease of changing how a platform allocates value, combined with the nature of a "two-sided market," where a platform sits between buyers and sellers, holding each hostage to the other, raking off an ever-larger share of the value that passes between them." Why do they do it? **Because they can**. Over time, however, this practice can create insurmountable problems.

The Spread of Unproductive, Company-Harming Products

Some observers argue that the idea of a "BS" product is offensive and unappealing. I use this term to encourage people not to overlook the issue. By labeling certain products as BS, I aim to highlight their minimal value and, at times, harmful impact on the business unit's performance, especially in the long run.

The issue is that political correctness regarding innovation and product development has taken hold across many organizations. It's when companies celebrate every product development. If a product successfully passes through the launch phase (somebody buys it), leaders are expected to declare it a wonderful success, and those who contributed to its creation and development deserve public praise. It's not until the product is managed in the market that you hear murmurs of disappointment.

Accelerating change drives an exponential increase in adjacent possibilities and, by sheer numbers, raises the likelihood of producing BS products. Sadly, BS products consume resources and distract focus from creating and developing more impactful products.

The critical issue is that every new product drives development efforts in one direction or another. When these efforts have independent paths, they can create an erratic trajectory that squanders opportunities to generate compounding gains. We might introduce a new product, but it won't integrate with other products to yield greater benefits. We don't need or want a straight-line trajectory for our efforts. Instead, we require a coherent trajectory that is unified, adaptable to change, and capable of delivering compounding gains.

The ineffective BS product issue stems from a common paradigm that influences product innovation and management in many large companies. I refer to this paradigm as "single-product, orderly systems thinking." Throughout the book, I discuss how this thinking blocks cost-performance leverage and positioning advantages. Companies stuck in single-product thinking will struggle with accelerating change. In contrast, those that adopt multi-product thinking will develop the capability and capacity to compete in the Accelerating Change Race.

The reinvention of product innovation and management must intentionally shift the context in which we think and make decisions. It requires transforming our thinking to embrace multi-product, complex systems while deliberately minimizing and avoiding incoherent (BS) products.

The reinvention is not merely an incremental advancement of our practices; it's a much-needed overhaul of how we think about and manage the development of products and services. A new approach is urgently needed if companies and management teams want to keep pace with their competitors and customers in the Accelerating Change Race.

It's a Tough Message

I understand that my message may not align with what most product innovation and management experts, possibly including you, wish to hear. It contradicts the incrementalism on which we've built our careers. Acknowledging the need to shift your perspective on product innovation and management to keep pace with the accelerating change is unsettling. But your business risks falling significantly behind if your management team clings to the outdated single-product paradigm.

Many experts are vocal in their opinions that the Accelerating Change Race means managers must change their attitudinal "mindsets" and behaviors. They must be open-minded, strategic, and trustworthy. That's nice, but it doesn't address the core issues regarding work, decisions, information, and a coherent stream of strategic moves. The context of multi-product complex systems in which people must think is more important than their attitudes.

Assumed Inability to Perform and Transform

Managers and leaders who adhere to the outdated two-pronged approach presume that those managing existing products cannot revitalize established product lines. They consider existing product lines insufficient for harnessing significant new technologies and business models. This perspective is understandable in traditional product innovation and management. However, the assumptions guiding managers to this conclusion hinder a business's ability to adapt to change.

The problem is that many managers constrain their thinking with assumptions anchored in single-product thinking. Single-product thinking focuses on creating, developing, and launching one product at a time.[25] This way of thinking is pervasive.

Old-school single-product thinking is prevalent in academic research, business publications, and public presentations. Perhaps most concerning, single-product thinking is ingrained in our processes and practices. We even base our innovation and product development software systems on single-product thinking. It underlies nearly everything related to product innovation and management.

Context Matters

A reinvented product innovation and management approach must shift our thinking toward multi-product systems. This shift involves more than merely expanding perspectives to encompass many products instead of just one; it also entails orchestrating a set of products as a unified system. This fundamentally changes how we approach innovation and product management.

Changing the context is powerful. Stuart Kauffman, a theoretical biologist and researcher in complex systems, suggests that altering the context in which we perceive a system enhances the likelihood of identifying influential factors that drive the system positively.[1] The issue holding us back isn't the influence or driver; it's the context in which we apply our thinking.

When we shift our thinking from orderly systems to complex systems, our perspective on influencing forces and constraints unveils a new realm of opportunities. Therefore, product innovation and management must be transformed.

The shift to product lines as systems encourages thought processes that steer efforts toward the continued improvement of the compounding gains we want the line to deliver. This shift also enables teams to overcome the constraints imposed by their companies when driving efficiency into their core products.[26]

To exit the Negative Innovation Loop, we must change capabilities, capacities, and approaches at the product line level. Product Line Systems Management does that.

But Why Product Lines?

Some people might ask, "Why not push harder on building new business models or be more aggressive with new technologies and 'radical' innovations?" Won't these efforts work to keep pace in the Accelerating Change Race? No doubt, these efforts may be beneficial in certain circumstances. The issue is that they don't help nearly enough. The solution should involve products and services that engage with customers.

So, "Why should we focus on product lines to keep pace with and lead the Accelerating Change Race? Why not individual products, a product family, or an entire business?"

I answer these important questions from three perspectives.

→From a Strategic View

There's an important principle in strategy that many of us may have never learned: "Excellent strategies are indivisible." You can't divide well-thought-out strategies and expect stellar outcomes. I learned this from renowned strategist Roger Martin.[27] Here's a summary.

Strategy Indivisibility

In business, we can divide entities into smaller units. For example, corporations can be divided into business units and further divided into functional groups. We can also separate product families into product lines, which may then be subdivided into individual products. See Figure 2.1.

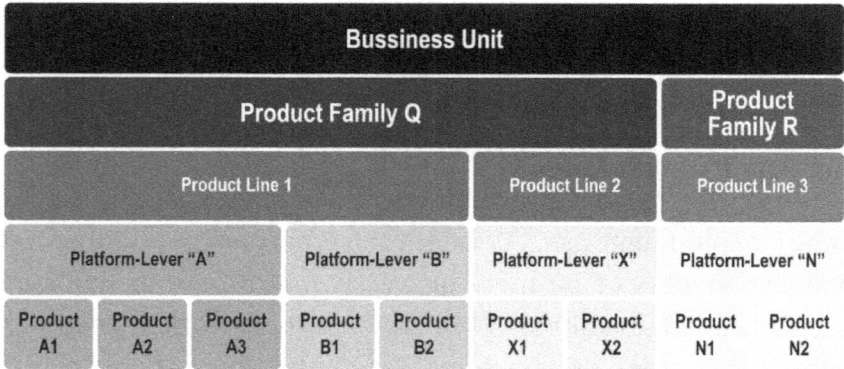

Figure 2.1. Product, Product Line, Product Family Taxonomy

Our question is simple: Which strategy domain yields the greatest gains in the Accelerating Change Race? Is each product the correct strategy domain, or could the product line, a product family, or perhaps the business unit serve as the appropriate strategy domain? The issue is that we forfeit potential gains when we excessively divide the strategy domain. We also lose focus on impactful elements when we select a level that is too broad. Both of these missteps reduce the strategy's effectiveness in addressing accelerating change.

Wrong Focus, Lost Gains

Two significant gains are lost if we concentrate on individual products instead of product lines.

First, we lose leverage, which enables faster development times and greater product performance per dollar. However, leverage requires multiple products and a range of customer needs. Individual products offer limited leverage, so significant volume is essential for a single product. While achieving leverage with a single product is possible, it is rare.

Leverage gains stem from two strategy principles that encourage companies to seek scale. The first is moving up the S-Curve to gain learning and development speed.[28] The second principle is Wright's Law[29], which states that cost will decline 15% to 20% as cumulative production volume doubles. In other words, scale improves each product's bang for the buck. However, notice that these gains focus primarily on the efficiency side of our two-prong challenge.

The second strategy gain we lose when focusing on individual products is complementary differentiation, which managers grew aware of through Harvard Professor Michael Porter's work.[30] In the consumer products arena, many companies heeded Howard Moskowitz's advice to extend differentiation.[31] It's why we have dozens of Coke flavors. Does anyone want Coke Marshmallow? Yes, that is a thing. The differentiation can boost the entire set of products.

Teams intentionally design the attributes of one product to complement those of another. However, this coordination must occur at the product line level. This advantage is not available to individual products.

The product line level allows for strategic moves that the single product level cannot achieve. Product line teams can intentionally position products in adjacent market segments to appeal to a broader customer base. The strategy poses significant challenges for competitors. While competitors may focus on one segment, they often struggle to address all segments simultaneously. Furthermore, such strategies enable the product line to achieve economies of scale, benefiting all its products. Competitors struggle to attain this scale, which compels them to accept a smaller market share and redirect their focus to other markets.

A key principle of product line management is targeting market segments with greater specificity and skill than competitors. Such targeting leads to higher customer satisfaction, obstructs competitors, and enhances cash flow. The crucial insight is that we lose these advantages if we concentrate on a single product instead of a product line.

Keeping Other Gains

But why not focus on product families or at the business unit level? Both product families and business units can contain several product lines. See Figure 4.1. The logic supporting a focus on product lines rather than product families follows the same principle. By selecting the product line, you also retain the gains and benefits of the product family and business unit.

Problems arise when you shift your focus to a product family that includes multiple product lines. This broader approach diverts attention from each individual product line without yielding additional benefits. Product line management can operate independently while remaining aligned with the management of a product family and benefiting from its overall strategy. Consider how Apple's iBrand strategy encompasses the entire product family, while each product line team (phones, computers, and watches) can progress as necessary.

The same applies to the relationship between the product line, a business unit, and the line's strategy. Apple's retail stores, intellectual property, and supply chain strategies align with product lines but do not control them. The greatest gains come from each line's efforts to enhance its products. From there, we seek gains only achievable across lines.

→From a Systems View

We reach the same conclusion when we answer, "Why product lines?" from a systems perspective. However, we frame it in different terms. Instead, we would argue that the product line system "substrate" provides the best orientation to address the Accelerating Change Race.

In systems thinking, the substrate refers to the area of the universe that you intend to examine. It establishes the context for your reflections. (See Appendix 3.) More importantly, this context is shaped by the constraints and forces that you must navigate. Plus, to maximize our influence over these forces and constraints, we must engage with a system at its "finest level of granularity."[32] I address this important topic throughout several ensuing chapters.

To determine the best system substrate for addressing the Accelerating Change Race, we need to evaluate each option to see which accounts for a sufficient range of forces and constraints while providing the finest level of granularity in system components and agents that influence outcomes.

Consider all components and agents within an entire product line system. In systems management, we strive to understand and influence their interactions. The product line system, rather than an individual product, serves as the foundation that enables teams to function at the finest level of granularity.

A challenge for systems experts is that they may understand the granularity principle but do not recognize or comprehend all current and potential parts and agents—the granules within the system. Business experts face the opposite challenge; they may understand the system's parts, but they do not adhere to the granularity principle or other system concerns. If you are unaware of the system's parts and agents, managing the system to achieve the best outcomes and improve the line's performance will be problematic.

The business and product family levels represent systems that are overly broad, with granularity too large to drive significant performance improvements. Individual products are smaller systems but face the opposite challenge. Focusing on one product at a time does not provide the cross-product interaction necessary for strategic maneuvers related to multiple market segments and leverage.

When examining the various product levels from a systems perspective, we reach the same conclusion regarding an indivisible strategy. The product line is the best system substrate for overcoming the efficiency-change dilemma.

The concept of the finest level of granularity is most impactful for the change aspect of the efficient change dilemma. Recognizing that a change to a product line, for example, or any strategic move, including scaling, must be accompanied by adjustments to the system's parts and agents (the granules), allows us to modify a business's core product. This is what I refer to as the Product Line System Change Theory.

The theory involves applying principles of complexity science to product lines as complex systems. Chapter 21 dives deep into Product Line System Change Theory, and Appendix 2 further discusses product lines as complex system substrates.

→From a Practical View

In practice, no mid-sized or large company offers just one product for sale. Even a firm selling a commodity will provide variants of the item in terms of form, shape, or packaging. This variety exists because customers need choices. They have different wants, desires, and needs. The outcomes that customers seek from products vary, creating distinct market segments and requiring unique products.

Granted, companies often assign the title of "Product Manager"— as in the singular product—to those who manage products. It doesn't mean they manage only one SKU or a product with no variants. Nowadays, the singular "product" title may be assigned to anyone who manages one or more products.

Large companies often manage multiple products simultaneously, while mid-level managers oversee several of these products. A disconnect arises when cross-organizational communication focuses on only one product at a time or addresses resource issues instead of strategic initiatives. This mismatch can lead to complications. I address this issue in Chapter 5.

From an operational viewpoint, the most important improvements in running the Accelerating Change Race take place at the product line level, rather than at the individual product, product family, or business unit levels.

Product Lines are Key

Product lines are among the least studied yet most impactful topics in business, significantly influencing the well-being of medium and large-sized companies. However, tactics, strategies, and approaches related to product lines are often overlooked in a company's efforts to increase sales and reduce costs.

Product lines are bespoke business entities that need management to address their issues and capitalize on opportunities. As I discuss throughout the book, product lines provide value gains that accrue as improved customer satisfaction, competitiveness, and cash flow. Companies realize these gains only by focusing on and acting on product lines.

Big companies must greatly dampen their old-school, single-product thinking and approach to innovation and product management, replacing it with a strong emphasis on product lines as multi-product systems. This shift minimizes wasted efforts on BS products and maximizes the cumulative benefits that can accrue across multiple products.

Product Lines and the Accelerating Change Race

Focusing on product lines, rather than individual products, is the most effective way to create and execute strategic moves that enhance product performance. It offers system parts and agents at their lowest level of granularity, allowing for maximum influence over outcomes.

The product line is also the lowest level at which a strategy is indivisible. You won't realize similar strategic gains by managing at the lower single-product level. That's why teams create, produce, and market products as part of larger product lines.

Product lines allow companies to deliver and change products more easily. Seeing a product line as a system also enables companies to build strong yet changeable product offerings.

Teams must oversee a line's strategy and influence its outcomes to gain efficiency and changeability. Product Line Systems Management helps teams do this. It allows teams to adjust and modify the system's components at their most detailed level.

Chapter 2 Summary

Why Product Lines?

- **The Organizational Approach:** Experts say the efficiency change dilemma is rooted in organizational issues. They attribute the problem to leadership, mindset, and culture. This is misguided.

- **Product Line Perspective:** Changes only to organizational approaches will not solve the problem. We must also change how we create, manage, and transform products.

- **Assumptions about Product Management:** The belief that product management cannot transform established product lines hampers our ability to adapt to new technologies and business models.

- **Why Product Lines?:** The strategic and systems views highlight how product lines offer notable, often unrealized, advantages. Product line management affords greater bang-for-the-buck performance (leverage) than single-product management. It also enables a synergistic boost on one product from the others. Plus, it sets up more powerful strategic moves to drive efficiency and induce change.

  1. **From a Strategic View:** Excellent strategies are indivisible. An approach focusing on individual products instead of product lines loses economies of scale and cross-product performance gains. The Product Line System is the most impactful for boosting performance and inducing change.

  2. **From a Systems View:** Focusing on the finest granularity level of a system is crucial for gaining the greatest influence on the system's outcomes. Managing the product line system's parts, rather than individual products, offers

influence at the finest level of granularity and leads to more impactful results.

3. **From a Practical View:** Large companies seldom sell and service just one product or service.

Chapter 3

Creating Value

Merging Thought Worlds to Perform and Transform

Value plays a critical role in the Accelerating Change Race. When a manager or project contributor mentions "value," they're likely referring to the product's value stream, which represents the value gained at each step of product development. The value stream approach establishes a solid foundation for product line management, but we must advance it further.

In product line management, we aim to capture value measures that help us address progress toward efficiency and change. Managing value for one while neglecting the other is unproductive. Our goal is to assess the value of both and do so in the near and long term.

Multi-Dimensional Value

Value in a product line context isn't one-dimensional. It's not just about money. Money alone might be beneficial if we wanted only to drive efficiency. However, when we need to change, two other dimensions of value come into play—customer satisfaction and competitiveness.

Most product managers who follow traditional practices will highlight customer satisfaction and competitiveness in their annual objectives. However, they do not connect these factors with value chains. While these factors are important, they are not operationalized. In product line management, all three factors must take center stage.

Let's examine the three factors before exploring their role as the most critical performance indicator for a product line and how they assist teams in addressing accelerated change.

• **Cash Flow**

Financial analysts often describe businesses in terms of cash flow and suggest the impact on equity valuation. It's also common for stock analysts to examine and report on efficiency challenges related to near-term cash flow. Unfortunately, if a CEO agrees with this perspective, product managers must lean towards efficiency and away from change, reinforcing the Negative Innovation Loop.

A problem can arise when a company keeps significant product changes separate from its core business. This creates additional pressure on the core business to generate even greater positive cash flow. Core products become the company's "cash cow." However, efficiency gains stagnate as products mature and sales decline. This leads analysts to cite poor cash flow as evidence of equity value degradation and to advocate for further cost-saving and efficiency-building measures, rather than focusing on core business growth.

Still, what happens to a core business does not improve the chances that Corporate Innovation will successfully scale a new business opportunity. Companies cannot rely on growth from a new business to compensate for a decline in a core product line.

In day-to-day operations, some old-school managers cling to the mantra that "cash is king." While there is truth to this guidance, it doesn't capture the whole picture. If you let cash become the ruling monarch of a business, it's unlikely you'll succeed in the future, especially as the pace of technological and market changes quickens.

• Competitiveness

Warren Buffett is well known for incorporating a competitiveness measure when evaluating businesses. He refers to this measure as a company's "competitive moat." The moat represents the firm's ability to defend against attacks from competitors on that company's "cash flow castle." Yes, he is very interested in cash flow. However, Buffett also aims to understand a firm's ability to compete while maintaining or growing its margins. Today, you'll find that the stock analysis firm Morningstar sells moat analyses on publicly traded companies.[33]

Lacking a competitive moat and being unable to establish one was the central theme of a Google engineer's internal memo regarding the firm's response to OpenAI's ChatGPT market entry.[34]

He and his team stated they could not identify a "competitive moat" for Google in artificial intelligence-enabled search that would continue to support their excellent pay-per-click advertising engine. This is undoubtedly a significant issue for Google. Their moat—the competitiveness of the search engine—is crucial to the business's performance, as it is for all product lines.

Product line competitiveness is not a winner-take-all or kill-or-be-killed challenge. Instead, it involves delivering value to customers in a way that influences competitors away from your domain. The greater your line's competitiveness, the more influence it exerts on competitors. In the Warren Buffett analogy, it's this influencing effect that forms the moat.

- **Customer Satisfaction**

Competitiveness and cash flow aren't the only concerns for innovators and developers in large companies. Their goal is to solve problems and ensure that the solutions delight customers. For innovators within big companies, customer satisfaction is the top priority. Identifying and solving customer issues is their primary responsibility. Managing cash flow and competitiveness is secondary.

The value of customer-centric innovation is well-researched and documented.[35] Customer experience (CX) and user experience (UX) concerns are mostly single-product, not multi-product. For product lines, overall customer satisfaction is crucial to their performance.

A product line team must understand current customer satisfaction and anticipate its direction for the future. That's a tough job. Satisfaction largely, but not entirely, depends on how well tangible and intangible product attributes match customer needs, wants, and desires. Both sides of this match can be uncertain.

If the team lacks insight into customer satisfaction's direction or changes, they'll remain unaware of this critical performance measure. Instead, they'll rely on fortune to enhance their line's performance. As you can imagine, this is not the smartest approach. The solution is to build knowledge and gain a better understanding of satisfaction by identifying the "jobs-to-be-done" that customers aim to fulfill. I address this in several chapters. In Appendix 7, I explore Jobs-to-be-Done analysis for product line attributes positioning.

Product Line Outcome Velocity: The Outcome-Driven KPI

Teams implement the three factors by plotting their values on a three-dimensional graph. This establishes a single reference point for managing the product line. However, in product line systems management, we can also generate similar reference points for different times, such as the past, present, and future.

Teams create a vector by connecting two reference points. Like all vectors, it possesses both length and direction. This vector represents the magnitude and direction of changes in product line performance, whether from the past to the present or from the present to a future point in time. I refer to this vector as the Product Line Outcome Velocity.

In the next chapter, you'll read how we create value in product lines is directly related to the Outcome Velocity vector. The vector is the product line's most critical performance indicator and is crucial to keeping pace with the Accelerating Change Race.

Creating Value and the Accelerating Change Race

Managers who believe mature product lines can't change will support dividing innovation from core operations. However, this belief is wrong.

Teams begin the approach to overcoming constraints to a line's changes and advancements by focusing on three value-creating factors: customer satisfaction, cash flow, and competitiveness. Changes in these factors combine to create the product line's "Outcome Velocity vector." This vector is a key indicator of the system's performance and change.

In the next chapter, I discuss the critical importance of the product line Outcome Velocity vector in product line management.

Chapter 3 Summary

Creating Value

- **Critical Role of Value:** Understanding value is essential for advancing a product line. In product line management, value is about 1. Cash Flow, 2. Customer Satisfaction, and 3. Competitiveness. When combined, these create a point-in-time measure of a product line.

  1. **Cash Flow:** Financial analysts often focus on cash flow, which, if prioritized for short-term gains, reinforces the Negative Innovation Loop, affecting a company's ability to adapt and innovate. Nonetheless, it remains a crucial measure of a product line's performance.

  2. **Competitiveness**: Competitiveness is essential to a line's performance. It reflects a company's ability to advance, adapt, or transform its product lines and create a 'competitive moat' to protect against attacks on cash flows.

  3. **Customer Satisfaction:** All product managers should aim to increase customer satisfaction with their products or services. A decline in satisfaction across a product line can be a problem. It may indicate that the product line is facing serious challenges.

- **Past, Current, and Projected:** Teams may calculate the scalar values of the three dimensions for different points in time. These values may be past, current, or projected (Delphi-futurecast: Appendix 4).

- **Product Line KPI Vector:** Connecting two different time points (past to current, current to future) creates a vector, revealing the value change. The next chapter explores the vector as a key performance indicator (KPI) for Product lines.

Chapter 4

The Product Line Outcome Velocity Vector

Influencing Outcome Emergence: Direction over Endpoint

T he speed at which teams develop new products has been a critical concern for over three decades. This focus has spanned the careers of most senior managers and executives. Today, however, speed is not our only concern.

The Time-to-Market Quagmire

Many consultants assert that expedited projects are crucial to the success of new products. Today, every major company utilizes at least one time-to-market key performance indicator (KPI) to enhance product performance.

The correlation between development speed and favorable outcomes has been undeniable.[36] But a big change has taken place. Nearly every company stresses time-to-market as a measure of development project speed. Consequently, the correlation between speed and profit is no longer as strong as it once was.

Product line teams now realize an added problem. They find themselves in an absurd Red Queen's Race, like that described by Lewis Carroll in his book "Through the Looking Glass" (Alice in Wonderland). To quote Carroll,

"Now here, you see, you must run twice as fast as you can to stay in the same place."

The problem is that development speed doesn't always enable a competitive advantage. Rather, it appears to work only when competitors are slow. Now, three decades into the time-to-market revolution, that's far less likely.

Outcome Velocity: Magnitude and Direction

In the dynamic, multi-product world, direction—not just speed—matters. For product lines, the direction toward improving outcomes is the most critical aspect. Teams must enhance cash flow, competitive positioning, and customer satisfaction across their entire product line, not just for one product. These three factors drive value creation and should be the primary focus of product line management.

The Outcome Velocity vector reflects a product line's performance. It is not a scalar indicator like the metrics used by most companies.

A vector is specific to a timeframe and has a direction and length that reflect the changes in the three dimensions during that timeframe. Teams can also add vectors to one another to show changes over multiple timeframes, and viewing this series helps teams understand how the line's performance is changing.

Consider how Amazon purposely changed its vector. For years, the company concentrated solely on competitiveness and customer satisfaction, without emphasizing profitability. This approach even earned them the tongue-in-cheek title of the world's most valuable "non-profit" organization.

In the three-axis vector graph, Amazon's Outcome Velocity scraped the bottom of the cash flow coordinates, while customer satisfaction and competitiveness values soared. More recently, Amazon intentionally changed the direction of the Outcome Velocity vector. They now drive cash flow while maintaining or improving competitiveness and customer satisfaction.

It is important to recognize that changes in product lines happen across three areas—cash, customer satisfaction, and competitiveness. This insight indicates that an Outcome Velocity vector is vital for guiding work and decisions toward efficiency and change. We also understand that when managing complex systems, a vector, rather than a scalar measure, is a better indicator of performance.

Time-to-Market versus Product Line Outcome Velocity

Faster time-to-market for individual projects has driven the Outcome Velocity of a product line for several decades. Historically, faster development project speed has correlated with improved Outcome Velocity. However, as I mentioned, the relationship between project speed and product line Outcome Velocity has diminished. See Figure 4.1.

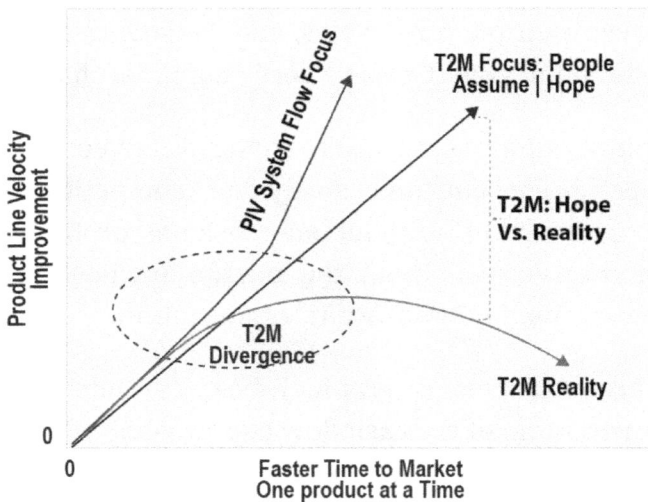

Figure 4.1. Time-to-Market versus Product Line Outcome Velocity. *In the past, Time-to-Market for development projects was a driver of a Product Line's Outcome Velocity. That's no longer the case.*

The two measures diverged as every company (including competitors) accelerated the speed at which they developed and launched products. Speed is a given in nearly every organization; direction is not. It's common to see competitors make strategic moves that undermine any gains achieved through faster project execution. For instance, Tesla's strategic maneuvers in the electric vehicle industry have rendered Ford's emphasis on project speed irrelevant. The issue isn't speed; it's the direction and magnitude of change that matter.

Notice how a product line's Outcome Velocity reflects aspects of both efficiency and change. If efficiency declines, near-term cash flow will also decrease; the same applies to customer satisfaction and competitiveness. Also, if the line's offerings remain unchanged, the cash flow may appear satisfactory in the short term, but competitiveness and customer satisfaction may diminish in the long run.

Once again, understanding a product line's Outcome Velocity is crucial to keeping pace in the Accelerating Change Race.

Improving the Outcome Velocity

Teams must deliver a stream of impactful outcomes to enhance the Outcome Velocity of a product line. The challenge for most companies is that their thinking and actions toward products often hold them back. Relying on outdated methods and practices is unlikely to be beneficial. Instead, teams and organizations must reshape their approach to products, systems, and flows to achieve significant gains. See Figure 4.2.

First, the context in which people think about product innovation and management is crucial. To enhance a line's performance, its team must shift from a single-product focus to a multi-product approach. Second, teams should revise their practices and models to align with complex systems rather than just orderly systems. Third, product line teams must incorporate information, decision, and outcome flows into their emphasis on workflows.

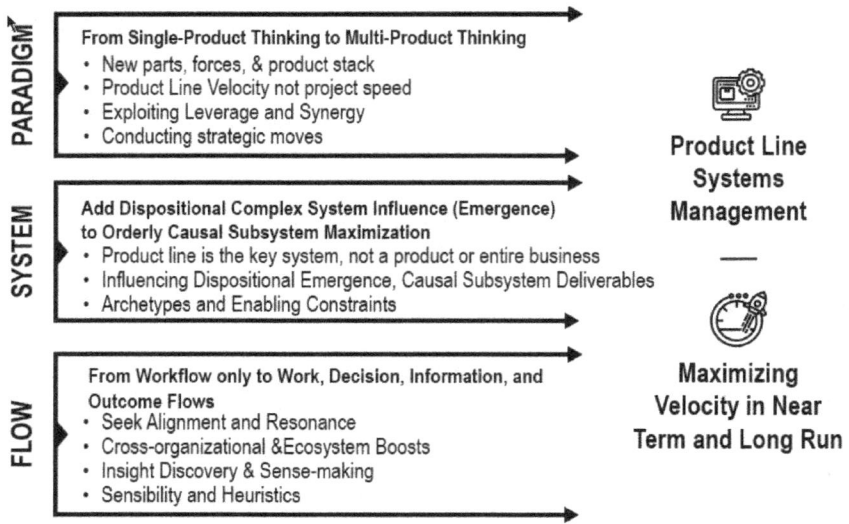

PARADIGM

From Single-Product Thinking to Multi-Product Thinking
- New parts, forces, & product stack
- Product Line Velocity not project speed
- Exploiting Leverage and Synergy
- Conducting strategic moves

SYSTEM

Add Dispositional Complex System Influence (Emergence) to Orderly Causal Subsystem Maximization
- Product line is the key system, not a product or entire business
- Influencing Dispositional Emergence, Causal Subsystem Deliverables
- Archetypes and Enabling Constraints

FLOW

From Workflow only to Work, Decision, Information, and Outcome Flows
- Seek Alignment and Resonance
- Cross-organizational &Ecosystem Boosts
- Insight Discovery & Sense-making
- Sensibility and Heuristics

Product Line Systems Management

———

Maximizing Velocity in Near Term and Long Run

Figure 4.2 Product line systems management requires change across three notable topics.

These changes affect how most managers perceive innovation and product development. The new approach shifts perspectives from a linear cause-and-effect view to one based on complexity theory. Flow introduces considerations of decision-making, information, and outcome flows into every company's workflow focus. You'll also notice that the multi-product context changes managers' priorities. It directs teams to focus on the components and agents of the product line system to create a coherent stream of strategic moves.

Focusing on improving a product line's Outcome Velocity transforms product innovation and management. It requires teams to advance their lines in ways they may not consider when operating in a single-product thinking mode and believing that cash flow is the only factor that matters.

The Capacity to Perform and Transform

Understanding how products and agents function as a system is crucial for effective product line management. This insight establishes the foundation for how teams manage the Accelerating Change Race.

The approaches and tools we implement for managing product lines as systems intentionally focus on enhancing a line's Outcome Velocity as a measure of performance. We also support a product line's change or transformation by increasing its Change Capacity. Outcome Velocity and Change Capacity work in tandem.

A line's Change Capacity refers to the amount and speed of change that the line can endure before negatively impacting the Outcome Velocity. This is why Outcome Velocity and Change Capacity are closely linked. As you'll learn throughout the book, Change Capacity is an issue related to people, systems, and assets.

I address both Outcome Velocity and Change Capacity throughout the remaining chapters. In Chapter 22, I delve deeper into change by discussing the Product Line Change Theory and explaining why managing a product line as a system is crucial to addressing the change aspect of the dilemma. Appendix 5 examines the nuances and challenges of proactively managing a line's Change Capacity, rather than relying on "after-the-fact" resiliency to overcome change issues.

Outcome Velocity and the Accelerating Change Race

Big companies must perform and transform. Addressing both is the most crucial issue these companies must tackle to stay in the Accelerating Change Race. To succeed, companies need to alter the context in which they develop products and foster innovation.

Switching the KPI to a vector helps alter the work context. The standard time-to-market metric does not address the need for simultaneous efficiency and change in the Accelerating Change Race. The vector reveals shifts in cash flow, customer satisfaction, and competitiveness. A change in these factors indicates a line's efficiency and capacity for change.

Chapter 4 Summary

The Product Line Outcome Velocity Vector

- **Three Decades of Project Speed:** Product development speed (time-to-market as a KPI) has been a significant concern for over three decades. The correlation between development speed and a product line's outcomes has always been strong. Now, as all companies emphasize faster time-to-market, that's no longer true. The universal focus on speed has led to a divergence of time-to-market values from performance results.

- **Absurd Red Queen's Race:** Product line teams face a challenge like the Alice in Wonderland "Red Queen's Race." Emphasizing project speed is so common that each company must "run twice as fast as they can" just to stay in the same place. Speed no longer guarantees an advantage.

- **Time-to-Market versus Outcome Velocity:** In the multi-product world, our focus shifts away from working faster and toward improving a product line's Outcome Velocity. It encourages and promotes better cash flow, competitive positioning, and customer satisfaction across the product line.

- **Improving Outcome Velocity:** Continuously improving a Product Line's Outcome Velocity demands

  1. A shift from single-product to multi-product thinking

  2. Adapting practices to a product line system's complex and dispositional nature

  3. Integrating work, information, decision, and outcome flows.

- **Active Structure, Processes, and Leadership:** Managing product line changes alongside organizational changes is vital for continuous improvement. Such a combined change must be the norm in product line management. It requires new leadership skills and adaptation of the organizational structure.

SECTION TWO

The Multi-Product Context

"Multiplicity, which is not reduced to unity, is confusion. Unity, which does not depend on multiplicity, is tyranny."
- Blaise Pascal, circa 1660; Mathematician, Physicist, Inventor, Philosopher, and Catholic writer.

Single-product thinking should be relegated to the entrepreneur's garage.

- The Author

Chapter 5

From Single to Multi-Product Thinking

Shifting The Context

Teams forfeit significant opportunities when they become trapped in a single-product mode. Context is important; unfortunately, the single-product paradigm in which most managers think and act is no longer as helpful as it once was.

Single-Product Context

Companies focusing on one product at a time encounter a challenge: This context shapes managers' thinking and is ingrained in their processes, templates, and methods. It also influences the rewards and accolades that drive behaviors. Most concerning, however, is that single-product thinking doesn't align with the multi-product landscape present in every industry.

Single-product thinking stems from the notion that one bundle of product features resolves against one bundle of customer needs. It leads us to believe we'll make more money if we create more products and resolve more customer needs.

Unfortunately, the idea that 'more products = more money' has its limits. It's ineffective in addressing the new efficiency-change challenge.

Old-School Shortcomings

Focusing solely on one product overlooks key opportunities to achieve cross-product advantages and wastes valuable efforts and resources. The troubling aspect is that many managers are perfectly content in this limited thought world because they don't realize that they are missing significant opportunities. As George Meyer, a writer for The Simpsons, said, "The most oblivious people are often the happiest." Unfortunately, in business, change inevitably devours oblivious happiness. The accelerating change we face today will obliterate it.

The missed opportunities caused by single-product thinking are significant. For instance, when you focus solely on one product, you are unlikely to design features that allow products to work together and create synergistic gains. Nor does this narrow perspective promote an interconnected ecosystem of products that minimizes competition. Instead, single-product thinking pushes companies into a reckless race to develop the next great product before their competitors do.

When single-product thinking dominates, the speed and efficiency of individual products, not the Outcome Velocity of the entire line, are the primary drivers of behavior.

New School Multi-Product Context

For decades, companies have learned to manage multiple products. You can see "multi-product thinking" in how they manage project portfolios and build roadmaps for future products. Still, most companies concentrate their efforts on the single-product paradigm.

I hope to convey a different perspective and, by doing so, help people realize that outdated, single-product thinking is no longer sufficient in today's rapidly changing world. Fortunately, we can learn from those companies that have made significant progress.

Apple and Its Chips

Consider Apple's strategic moves. It took years, but they introduced the M-series chips for Mac computers and iPad tablets, as well as the A-series chips for iPhones. Plus, they've tightly integrated macOS with the M-series chips and iOS with the A-series chips. These are not isolated product developments. Instead, these technological advancements represent major strategic moves that achieve notable leverage (or "bang for the buck") across the product lines. In the multi-product landscape, we refer to these types of strategic moves as platform-lever advancements.

The development of M-series and A-series chips impacts many of Apple's products and a wide range of apps available in the App Store. Apple is also making platform-lever moves for the Apple Watch (S-series chips) and its Vision Pro (R-series chips). What is crucial for Apple is the advantage the company gains across multiple offerings in various market segments and across several product generations. Apple does not operate within the outdated single-product model.

A Bigger Game

Apple understands that while rapidly developing products is important, fast development times are not enough. Instead, Apple aims to improve its products through numerous new versions continually. Their vision extends well beyond a single product. It requires much more engineering and marketing strategies than what a single-product scenario permits.

Apple's actions underscore an important principle at the core of multi-product thinking. No company should expect to compete in today's markets with just one product or in just one segment. You can't expect to win continually by going faster, one development at a time.

Product Lines, Not Products

Companies gain significant advantages by engineering product features to meet the unique needs of customers across various market segments. This intentional stream of innovations helps maximize customer satisfaction, thwart competitors, and enhance cash flow. It improves the velocity of the product line's outcomes.

The approach requires that tasks, decisions, and information—the flows within a product line system—operate in unison with a focus on enhancing the entire line of products, rather than just one product. Companies that excel at managing multiple products are better positioned to grow their brands, platforms, and customer support services.

Multi-Product Systems and Outcome Velocity

Understanding the components of a product line is fundamental to effective product line management. Each product, technology, and market segment functions as part of a system. While the elements that comprise a single product contribute to the line, the product line system includes other agents not in the makeup of individual products.

Notice how Apple's deliberate chip and operating system advancements change each of their product line systems while keeping their larger product family seamlessly integrated. Such moves don't come from old-school single-product thinking.

The multi-product mission aims to enhance the velocity of a line's outcomes. This is an ongoing task, not a 'one-and-done' project.

Comparing The Single and Multi-Product Contexts

The difference between considering one product versus multiple products is not merely a black-and-white distinction. There can be many shades of gray. However, you'll find that managers, departments, and entire companies tend to lean more toward one context and the thinking it induces.

Table 5.1 below outlines some differences between the single-product context and the multi-product context. If your company focuses more on one or multiple products, it's appropriate to describe it that way. For example, Apple sometimes has a single-product orientation, but the company also considers multiple products, with the multi-product context prevailing.

The opposite is true for many large companies. Managers in these companies who wish to improve their product line's performance must change their thinking.

Table 5.1 Single-Product compared to Multi-Product Thinking

This is not a binary choice; rather, it's a comparison of ends of a continuum.

	Single-Product	Multi-Product
Orientation	A focus on one product or one concept at a time.	A focus on sets of products, most often a product line.
Future View	Seeking the next great product... Relying on creative discovery, intrapreneurs, and lean startups.	A focus on advancing the full product line as a system and using systems thinking and creative thinking to form effective strategies.
Portfolio	Portfolio Management covers development projects only, seeking to optimize resource use.	Portfolio Management at the product line level encompasses front-end projects and other product line system projects, aiming to achieve greater Product Line Outcome Velocity and Change Capacity.
Development Focus	Oriented toward the fast development of an MVP (minimum viable product)	Focused on maximizing the product line Outcome Velocity.
Concepts	Front-end concept generation defines one potential product at a time, with loose targeting.	Front-end driven by coherent Jobs-to-be-Done market segmentation; purposeful concept targeting, purposeful strategy moves.
Speed vs. Outcome Velocity	A focus on a single product's "Time-to-Market."	Focus on Product Line Outcome Velocity - Continually seeking improvement.

Spotting Your Context

Examine the practices within your company or any other organization. Is the company oriented toward the single or the multi-product paradigm? Do the practices, methods, templates, and IT systems the company uses seek to advance one product at a time, or do they seek to gain the most from many products concurrently?

Recognizing single-product practices within your company is the first step forward. If you observe traditional single-product thinking, as I expect, can you also identify how it affects the velocity of your product line's outcomes?

Some managers refer to certain single-product practices as "best-in-class," which may prevent them from recognizing the issues these practices create. Instead, they regard studies on "best practices," particularly those from Harvard, Stanford, or McKinsey & Company, as sacred and not to be questioned.

There's an inherent problem with the approach many consultants and academics take toward best practices. They base their assessments—the questions and evaluation techniques—on the single-product paradigm. They measure how best to develop and manage one product at a time, not entire product lines. Then, they take this single-product understanding and extrapolate it to mean "do more of it." Managers interpret the extrapolation as a "best practice" and apply it across their product innovation and management initiatives. This approach, however, is a mistake.

To understand how the single-product paradigm influences product innovation and management, please take a few minutes to explore this concept on your own. Search the internet for blogs and articles on best practices related to topics like product discovery, user experience, or minimum viable products. You may also explore KPIs, agile, or waterfall development practices. Read the material and ask yourself whether it explains things from a single-product or multi-product perspective. The answer you'll find leans heavily toward single-product thinking. And that's a problem.

The Multi-Product Stack

Understanding the difference between single-product and multi-product thinking is one aspect; applying this knowledge is another. The transition occurs when we consider the product line from a systems perspective. To illustrate this difference, let's analyze the components of a product line.

Figure 5.1 shows two "stack" diagrams, one for a single product and the other for multiple products. Each diagram shows the key system parts that teams address in each paradigm.

Notice that the multi-product stack introduces parts that are not present in the single-product stack. These unique parts significantly alter our perspective on product lines. The traditional single-product stack undergoes notable changes and reflects a different system. This establishes a significant shift in context for innovation and management products.

Single-Product Stack	**Multi-Product Stack**

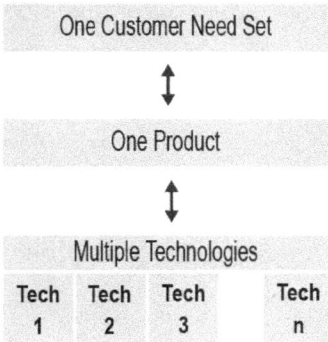

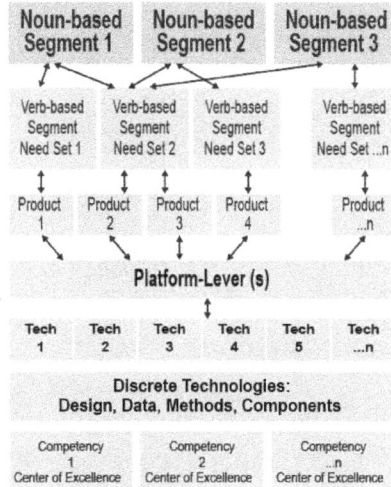

Figure 5.1. *Single Product Stack versus Multi-product Stack*

From a multi-product perspective, we're more focused on the entire product line offering than on just one product or system component. Each new product launch represents a strategic move that adds to the product line system. Ideally, each launch also enhances the overall outcomes of the system. However, new products aren't the only factors contributing to a line's gains.

Each current and potential component in the product line stack can enhance the line's performance. However, companies must shift their thinking and practices from a single-product to a multi-product context to realize these gains. This shift is essential for managing product line systems in our age of accelerating change.

Multi-Product Thinking and The Accelerating Change Race

Our concerns and expectations evolve when we consider multiple products rather than just one. We aim for strategic initiatives that benefit an entire product line, rather than simply launching each product quickly.

Multi-product thinking focuses teams on driving cash flow, beating competitors, and delivering greater customer satisfaction.

Shifting from considering one product to many is essential for keeping pace in the Accelerating Change Race. If a company continues to operate within a single-product context, it can only modify or enhance one product at a time, which is a recipe for failure.

Single-product thinking is problematic. Companies that adhere to it risk entering a negative innovation loop. Instead, they must embrace a multi-product approach to improve their lines' outcome velocities. These gains will help break from the pack in the Accelerating Change.

The transition from single-product to multi-product thinking necessitates a deep understanding of how product lines function as systems. Teams must identify what drives the system and learn how to influence changes that enhance the line's Outcome Velocity. In the next chapter, I lay out a foundation for these answers.

From Single to Multi-Product Thinking

- **Need for Multi-Product Systems:** Shifting from a single-product to a multi-product context creates significant opportunities and allows for more effective and impactful management of product lines.

- **Enabling Systems Management:** A systems approach to product line management requires a shift from single-product to multi-product thinking while considering both near- and long-term outcomes.

- **Multi-Product Context:** Shifting to multi-product thinking and practices is foundational to product line management. Multi-product thinking emphasizes cross-product leverage, intentional attribute positioning, and a longitudinal view—all necessary to improve a product line's Outcome Velocity and do it continuously.

Chapter 6

Product Lines as Complex Systems

From Orderly to Complex

Please don't let the word "complex" disinterest you in the topic. Throughout our careers, we've learned to keep things simple, and the word "complex" seems to suggest the opposite. Here, though, the word is in the title of a system type. It doesn't mean the system is impossible to understand or manage.

Appendix 2 explains the differences between complex and orderly systems. Understanding the difference is important.

Product lines are multi-product systems within a larger business system. More importantly, these groups of products are a business's most critical system. It's difficult to envision any large company without at least one successful product line. Similarly, it's hard to imagine a company not wanting to improve the Outcome Velocity of its product line.

However, product lines are complex rather than orderly systems. By their nature, they are adaptive and dispositional. We don't have a choice about whether a product line is a complex system; it simply is.

Complexity is Different

Don't be confused by the notion that a product line is a complex system. The confusion often arises from the fact that some parts within the product line system may be orderly subsystems. To some managers, the product line is merely a compilation of products. If they view these products as orderly systems, they assume the entire system must be orderly. However, that is not the case.[37]

First, product lines are not just a simple combination of multiple products. Second, most complex systems contain organized subsystems. Also, don't underestimate your ability to influence the outcomes of a complex system. Quite the opposite is true. What differs is that you need a deliberate approach to drive outcomes or, in the case of product lines, enhance the velocity of the line's outcomes.

The insight I wish to highlight is that orchestrating a product line as a complex system has a more significant impact than what is achieved by the orderly systems practices most companies employ to tackle straightforward challenges. To understand this, we must explore systems and complexity more thoroughly.

Systems Thinking

Many managers recognize Jay Forrester's insightful work on Systems Thinking from nearly 70 years ago. However, luminaries such as John Sterman, Russell Ackoff, and Peter Senge brought the topic to the forefront of modern business management during the

1980s and 1990s. You'll also find significant contributions from Lean and Theory of Constraints proponents like Taiichi Ohno, Edward Deming, and Eli Goldratt during the same period.

These renowned thinkers helped us understand how one thing can lead to another in social systems, such as businesses, schools, and governments. Even today, their presentations provide us with many "aha!" moments when we suddenly comprehend something in a new way.

We understand that a system exists as a whole. No part of the system operates independently to drive its full performance. The interactions among the parts are significant. Our concern should not be limited to just one part or force.

Still, the most common way people try to solve complex problems is by breaking them into smaller, easier-to-understand pieces and focusing on improving each piece individually. However, this technique contradicts what the experts taught us. Instead, we must prioritize improving the entire group of products over improving any single product. The whole system is more important than any of its parts.

The examples shared by the luminaries were straightforward and widely understood. Professor Ackoff explained that trying to create a high-performing car by combining the best engine, transmission, chassis, and so on from different vehicles is unlikely to produce the best car. The car may not even run. His point was that we need to view and create systems as complete entities. They are not merely the sum of their parts. We should adopt this perspective from the beginning, rather than treating it as a step along the journey.

Product Lines as Complex Systems

Product lines are no different. A team may think it's improving a line by adding one new product. They do everything possible to ensure the success of that one product. Systems experts refer to this as "optimizing the local." However, assigning resources to that product could under-resource other, much-needed strategic moves. Such mismanagement contributes to the BS products I discussed in Chapter 2.

Some managers may believe that this issue should be addressed by managing and prioritizing a set of projects or products. However, the challenge is that many strategic initiatives are not product development projects and do not align with that management style. For instance, enhancements to a shared platform-lever or significant market research should also be considered. Both could have a greater impact on a product line than the development of a single new product.

A key to improving a system's performance is to view the entire system while also recognizing that performance arises from the interaction of the system's components and enablers. This understanding establishes a critical principle in product line management. The primary objective is NOT to accelerate product development projects. Instead, it is to innovate and enhance the interaction among the system's existing and potential components, as well as other agents. The mission is to implement these innovations and advancements through a coherent series of strategic moves that increase the line's Outcome Velocity and enhance the line's Change Capacity.

Thinking about product lines as systems introduces us to several interesting topics. In this context, contributors create, modify, and occasionally remove parts of the system, mitigate constraints, and redirect forces. Through systems thinking, we make strategic decisions based on the interactions among all components and agents within the system. Individual products represent just one category of parts.

Product line managers and their teams must develop and utilize skills and knowledge related to both the current and future parts and enablers of the system if they wish to gain an advantage and enhance a line's Outcome Velocity. Systems thinking does not perform the work; it serves as a powerful tool to amplify the outcomes.

Dispositional Complex Systems

As I discussed earlier, many managers view systems as organized groups of components. They envision mechanical gears moving with near-perfect precision. At times, they'll suggest that the orderly system is complicated, and its parts may be separated by time and space.

In business, though, the orderly characterization is often wrong.

Most business systems are not orderly. They don't produce perfectly predictable outcomes. Instead, they are complex and dispositional. When one event occurs in the system, there's a likelihood—but no guarantee—that another event or outcome will happen. It's the difference between causation and correlation.[38]

Think of a complex system as a giant puzzle with many pieces. But here's the catch: we don't know if the pieces fit, and we're not 100% sure what the final picture will look like. Both the pieces and the picture can change, further complicating matters. It's like trying to predict the path of a hurricane—we can get a general idea of where it's headed, but we are unsure exactly where it will strike or the damage it will cause. That's why we refer to complex systems as "dispositional." Their outcomes aren't guaranteed, even if we have some ideas about their likelihood. They possess a propensity or disposition, not certainty, toward outcomes.

The advent of rapidly improving AI has created an intriguing juxtaposition of orderly and complex systems. You'll hear people claiming expertise in one topic or another declaring, "Use my AI tool to predict an outcome accurately." Such assertions may be valid for an orderly system but not for a complex one. The tool may help generate insights, but it won't deliver accurate predictions of outcomes. A system where an outcome can be accurately predicted is orderly, not complex. AI cannot transform a complex system into an orderly one. Along with machine learning, it may, however, move us on the continuum between these system types.

Ironically, efforts to predict complex systems can introduce rigidity that undermines their very nature. The problem arises when teams become overly reliant on orderly system-based methods. Attempting to pinpoint causes and effects can stifle the potential of complex systems to generate outcomes, effectively slowing down progress.

Uncertainty Changes Things

Complexity and uncertainty distinguish dispositional systems from orderly systems. This difference is significant when applying practices based on the principles of orderly systems' flow to enhance the outcomes of a complex system.

Consider how management techniques aimed at improving orderly systems seek the "root causes" of hindrances, issues, or problems. In complex systems, by definition, there is no such thing as a "root cause." Instead, managers must identify patterns in system behavior. This is because complex systems, including product lines, can consist of many parts and enablers interacting with numerous forces and constraints. The collective behavior of all components, rather than a single root cause, determines the system's outcomes. Recognizing and understanding the behavior patterns of a product line system is key to enhancing the line's Outcome Velocity. I dive deep into this notion in chapters 9 through 13.

Attempts to use root cause analysis (RCA) techniques and methods to identify and mitigate a single cause will be ineffective for complex systems. Their application encourages standardizing outcomes, which goes against promoting change. It hinders the emergence of outcomes and the development of Change Capacity within the line. However, this insight doesn't suggest that we should discard RCA methods such as the 5 Whys, Fishbone Diagrams, Pareto Charts, or perhaps Lean or Theory of Constraints. They hold too much value to be overlooked. Instead, we must enhance our ability to discern when and where to apply them—another crucial principle in product line systems management.

Adaptive and Self-Organizing

You'll also hear complex systems referred to as "complex adaptive systems," with some experts using the acronym CAS.[39] A complex adaptive system consists of parts and forces (the system agents) that adapt and self-organize without intervention or management. These agents will influence system behaviors that navigate around hindering forces to enable emergence.

Even without a team's understanding of their product line as a complex system, the product line will, to some extent, adapt to its surroundings. This impressive characteristic means product lines can survive without intentional systems management. Unfortunately, this survival doesn't always yield the best possible performance. This is what happens when companies lay off employees without altering their products and services.

The issue for product line management is that adaptation and self-organization do not guarantee the best outcomes or eliminate the possibility of the line's failure.

Most adapting behaviors for product lines will prioritize short-term cash flow gains over long-term competitiveness and customer satisfaction. Over time, this cash-focused short-sightedness will affect the line's Outcome Velocity, and not in a positive way.

Outcome Emergence from Complex Systems

The total outcome of a complex product line system does not emerge from just one part or agent. Instead, many, if not all, components must interact to produce outcomes. Consider the human body. Our body parts—organs, limbs, and brain—function independently, and the marvel of life comes from their interactions. It takes many parts and forces to make it work.

Moreover, the components and forces within complex systems are fluid and ever-changing; even subject matter experts do not always fully grasp them. A fitting analogy for a complex business system is a biological or environmental system, rather than a collection of mechanical gears. Complex systems resemble the Amazon Rainforest or viral evolution more than they do a Ford F-150 or a Rolex watch. Remember this when you see a presentation or read an article that incorrectly chooses a picture of a gear to represent innovation as a system. Gears represent orderly systems, yet product innovation, including development and management, is a complex system.

We Know Much More

For the past 50 years, scientists and experts have investigated the functioning of complex adaptive systems. They have helped us understand the various parts and forces that drive these systems. Now, we can identify complex systems and recognize many factors, both internal and external, that influence them. Even better, we can create models to illustrate how the components, forces, and outcomes are interconnected.

Machine learning is a powerful tool for unraveling the messiness of complex systems. It can identify connections among various components and agents, even though it can't tell us precisely why things happen as they do. This is positive news for managing these systems, but it doesn't imply that we can predict everything perfectly. There will always be uncertainty. Our role is to use our knowledge and make sense of situations, even if we're required to devise plans and take actions on the fly.

Complex Systems in Business

Wherever you look in business, complex systems are at work. These systems consist of many moving and evolving parts, which can be unpredictable. Consider how unexpected events—such as the pandemic, wars, or even bad weather—can derail operations. You'll also see gradual changes, such as consumer preferences, technological advancements, tariff threats, and the overall economy, can disrupt how a business operates.

We've learned that unexpected events can create significant problems. Remember COVID-19? It disrupted supply chains and altered consumer demand. Emerging technologies and changing customer behaviors can also cause upheaval. The key for businesses is to recognize that "stuff happens" and to be prepared to address the wide range of potential challenges that may arise.

When companies depend on orderly practices in complex adaptive systems, the outcomes can be unsettling. Uncertainties build up and drive the system toward chaos. We may perceive this as turbulence in a workflow or decision-making process, leaving contributors dissatisfied with the orderly systems approach. Nevertheless, those who still advocate for this method, including experts and evangelists, often cite research based on single-product orderly systems thinking. They contend that this approach is a proven best practice that should not be challenged.

In keeping with its complex and adaptive nature, the system will ultimately separate from such single-product, "best-in-class" practices. Evangelists can influence behaviors only to slow down the adaptation, not stop it.

The issue with self-adjusting (adaptive) change is that it can waste both time and money, resulting in delays in strategic moves. Unfortunately, these delays can have long-term negative consequences.

Timing and Impact

Company leaders who navigate significant challenges often refer to their organization's "resilience." This suggests they excel at adapting to unexpected changes and functioning as a complex system. They might say, "Our company is tough; we bounced back from that disaster!" However, there's more to the story. The speed of your response, the specific actions you take, and the extent of their impact are all critical.

Imagine you're facing a major "Black Swan" event. Almost any company can scramble and adjust. But if you wait too long to make a real plan, your situation can worsen. It's like holding back a major flood with a few sandbags. You might slow the water, but it will still flood through nonetheless.

Sure, it's important to be flexible and adaptable. However, strong leaders don't simply wait for trouble; they think ahead, make smart moves early, and guide the company in the right direction. Consider how Toyota has responded slowly to the electric car market. I mention Toyota because this respected company excels in orderly system methods. Yet, their slow response may have consequences that don't look promising, especially in markets where Chinese companies compete. Given the company's success, I can't help but wonder if I've overlooked a technology, market trend, or system behavior. Did I miss something that Toyota understood?

If you ask the question about Toyota on LinkedIn, you'll quickly notice the divide between traditional orderly system experts and modern complex system experts. The main issue in the comment dialogue is that orderly system thinkers often fail to recognize that complex systems are different from orderly systems. They tend to use their orderly systems hammer and try to treat the complex system as if it were just a collection of orderly system nails. But it is not.

Knowledge and Uncertainty: The Cynefin Framework

A valuable framework called Cynefin (pronounced kuh-NEV-in) is a game-changer for managing complex product lines. Dave Snowden, a renowned thinker in complexity theory and knowledge management, created Cynefin. This framework helps us understand the levels of uncertainty within a system. Based on these uncertainty levels, Cynefin provides insights into how the system's components interact and how outcomes emerge. While the full depth of Cynefin extends beyond this brief explanation, an essential element to grasp is this: our knowledge and the system's uncertainty dictate the best ways to manage it and influence the emergence of its outcomes.

We must manage complex systems differently from how we manage orderly systems.

Cynefin highlights why a heavy reliance on orderly cause-and-effect methods can often backfire in complex systems. It emphasizes that our level of knowledge plays a crucial role in managing a product line as a complex and ever-changing entity. The product line team must align with the level of uncertainty inherent in the product line system.

The Doubling-Down Efficiency Mistake

A significant issue with using simple cause-and-effect methods on complex product lines is that teams become entrenched in their efforts to produce the same products more quickly and at lower costs. Managers often embed this approach throughout the entire company. This reliance on cause-and-effect methods complicates future changes. Such a situation is counterproductive when leaders need to be efficient and adaptable in order to perform and transform.

The issue is that cause-and-effect methods treat the product line like a simple machine — an orderly system. We observe this with approaches like "Lean" and the Theory of Constraints. These methods are designed for more predictable situations, not for a complex dispositional product line system.

A team's goal shouldn't be to find one single reason for a problem or force everything to be the same. Using cause-and-effect methods on complex product lines can work against our goals by making the organization even more resistant to changes.

Focusing on simple cause-and-effect solutions for an entire product line might seem like a smart move at first, especially for managers who can't envision the need for change. However, maintaining the same approach for too long can be risky. If you doubt this, ask the former employees of the many companies that sought greater efficiency to avoid bankruptcy. They needed change, but they prioritized efficiency.

When a company becomes trapped in producing the same products, any changes can seem intimidating and threatening. This situation may cause teams and entire organizations to take too long to make critical decisions and implement meaningful actions. This is likely to lead to negative outcomes. If a company prioritizes efficiency over change, it may become nearly impossible to address the issues.

Changing Outcomes

The important takeaway for product line teams is that outcomes from complex product line systems must change for the system to survive. This principle is crucial for driving and improving the Outcome Velocity of a product line.

A product line team's secret weapon is building the capacity for change into the system without getting bogged down in orderly cause-and-effect thinking. I discuss this in many ensuing chapters.

Dave Snowden points out another critical concern when seeking to understand and advance a complex system. He explains that we must understand our system at its 'finest level of coherent granularity.'[40] The concept of granularity poses a significant challenge for many companies. Managers who are entrenched in single-product thinking will find it difficult to grasp the full spectrum of components and elements without a complex systems view of their product line. Their perception that the system is a single product is flawed; their lowest level of coherent granularity will be inadequate. In terms of complex systems, they are operating on the wrong substrate. See Chapter 2 and Appendix 3.

Complex Systems and the Accelerating Change Race

Complexity theory explains that no single root cause hinders the outcomes of a complex system. It also explains how and why such systems are adaptive and dispositional. Teams must extend beyond root cause methods to achieve exceptional product line outcomes.

Instead, we've learned that to drive outcomes toward efficiency and change, we need to explore and experiment with the system's parts and agents. It's their interactions that matter. This is a fundamental principle of complex systems thinking.

For product lines to evolve, teams must view them as complex systems. The role requires a deep understanding of the system's components, agents, constraints, forces, and their interactions.

There are ways to help the complex system move in the right direction. Teams can adjust their product line systems' assemblage and influence it toward better outcomes that yield greater Outcome Velocity.

Chapter 6 Summary

Product Lines as Systems

- **Dispositional Complex Systems:** Product line systems are dispositional and complex, not orderly. They lack predictable outcomes, and their parts and forces are more like biological or environmental systems than mechanical gears.

- **Misconceptions about Causality:** Some managers think product lines are orderly systems, assuming the system follows cause-and-effect principles. However, product lines are not just a roll-up of many products, with each product an orderly system. While product lines are complex systems that may contain orderly subsystems, there is no such thing as a single root cause of a complex system's hindrances. Those looking for it will waste their careers.

- **Influencing Complex Systems:** Teams can influence and guide a line's outcomes through a deliberate approach called Product Line Systems Management (PLSM). This approach focuses on continually improving the line's Outcome Velocity by purposely influencing the interaction of the system's parts. PLSM differs from the common practices used by many companies.

- **Systems Thinking Overview:** Systems thinking emphasizes the interdependence of system parts. However, the best parts from different systems will not create the best system. The parts must interact and work in unity. Teams must view their product line as a complete entity, not just the sum of many parts. When teams have this perspective of the product line, boosting the line's Outcome Velocity is much easier.

- **Outcome Emergence in Complex Systems:** The outcomes of a product line system emerge from the interaction of many parts.

Knowing and acting on just one part may be helpful, but it's insufficient to improve a product line's Outcome Velocity continuously.

- **Knowledge and Uncertainty – The Cynefin Framework:** The Cynefin framework helps us manage knowledge and uncertainty in complex systems like product lines. It highlights that our state of knowledge matters when dealing with a complex dispositional system.

- **Avoiding Efficiency Inertia:** Applying flow principles to complex systems but maintaining a single goal to deliver more of the same products leads to efficiency inertia. Delivering standard outputs faster hinders positive change and amplifies challenges in overcoming accelerated change.

Chapter 7

System Flows, Boosts, & Constraints

Work, Information, Decisions, and Outcomes

Wn we shift our thinking to managing product lines as complex systems, we also change how we see the system's flows. The concern is no longer only about speeding up and debottlenecking workflows. Instead, we're concerned about continuously improving the line's Outcome Velocity—the product line system's outcome emergence.

AI-induced accelerating change intensifies the demands on all system flows and their intertwined effects.

Spot The Flow

Flow, like 'Systems,' has been a popular descriptor of business practices for several decades. Flow comes from manufacturing and lean principles[41], which describe production throughput as a flow. Today, managers often describe their product creation process as a flow or a stream.

Most people interpret the word "flow" as a reference to fluids, gas, or electrons moving in a steady stream. But for product lines, the flow isn't laminar and steady. Instead, it is usually somewhat turbulent. And 'flow' does not relate only to completing projects or launching new products, as many managers believe.

It's Not Just Cash Flow and Workflow

Managers commonly apply flow principles to work, cash, and value creation. It seemed to be the perfect match for Lean methods because it introduces an outcome continuum. However, for product lines, we don't have straightforward streams. Instead, we pursue outcomes that aren't flowing until they emerge from our work. Our challenge is to enhance the emergence of a product line's outcomes. But achieving this requires us to improve all the flows in the line. Let's explore why.

When applying the flow analogy to product line management, we must address three concerns. First, teams need to understand and communicate what is flowing. Next, they must be able to implement changes that enhance the flow. Third, teams should learn to avoid impediments and forces that negatively impact the flow of outcomes.

Transferring flow principles to product lines is not a straightforward process. If a concept or idea is good, we'd likely incorporate it into our flow and use it. We wouldn't accumulate good ideas in a buffer or place them on an inventory rack. The significant challenge for most companies is that they lack the necessary methods and practices (enablers) to guide and enhance the emergence of a product line's ideas and concepts, as well as their outcome flow. Plus, they often lack the tools to minimize obstacles to such emergence and flows.

Four Distinct Flows

Product line system flows differ from those typically addressed by managers in innovation and product management practices. Here's a closer look at the flows within a product line system and how they impact one another.

• **Workflows**

Project managers easily understand workflows that are driven by waterfall techniques, gated processes, Gantt charts, and Agile methods. Numerous workflows exist in a product line system, and these flows are essential for completing tasks. Most development managers focus on finishing projects more quickly and delivering results faster, and using Time-to-Market serves as their primary performance indicator.

However, product line workflows are not solely about product development projects; non-development projects are also important. For instance, market research and strategy formation must be conducted continuously, even though neither qualifies as a development project. The emergence of a line's outcomes will be hindered when this work is executed poorly or neglected altogether.

You'll also notice that orderly system techniques may not be sufficient to enhance outcome flows. Consider how Lean and the Theory of Constraints help teams identify, mitigate, or eliminate the root causes of workflow obstructions and bottlenecks. These practices emphasize driving efficiency, rather than change, and are firmly rooted in orderly systems thinking. In product line management, these practices can improve workflow but may not address the product line system as a whole. While we strive for improved project work, our primary concern is the line's Outcome Velocity and Change Capacity.

• Information Flow

An information flow consists of documents, data, knowledge, and models. It is quite different from a workflow. At the core of this flow is information about customers, markets, technologies, and competition. This information serves as a basis for gaining insights into the interplay among the system's current and potential components and agents.

Each component and force in a product line system has associated data and information. Analytics, exploration, and observation can convert this data into valuable insights.

Information enables a system to function. Information flow enables organizations to capture, store, manipulate, and transform data effectively. The challenge lies in selecting software that best fits the product line system. Unfortunately, it can be painful when a software application doesn't align with the product line. From personal experience, I can say it's even more frustrating when a software vendor lacks understanding of multi-product systems.

One of the fascinating aspects of a product line's information flow is that new knowledge emerges from people's thinking, whether it is creative, analytical, or strategic. We refer to new knowledge that arises quickly as an insight. When knowledge develops over longer periods, we refer to it as learning.

The emergence of knowledge is a necessary precursor to the emergence of outcomes. Anything that obstructs or hinders insights and learning will also impede the emergence of outcomes. A key principle in product line management is that teams striving for improvements in Outcome Velocity must identify and minimize impediments and obstacles to knowledge building and insight generation. This contrasts with traditional product management, which focuses heavily on the speed of project workflows.

Product line teams need to generate insights and learn more quickly and consistently than competitors and new entrants to keep up in the race of accelerating change.

• **Decision Flow**

The decision flow represents the interconnected progression of choices and judgments. This flow requires coherence across all judgments, choices, and decisions, especially those that impact investments and resource allocation. Without coherence in these choices and judgments, decision flows cannot exist. Furthermore, as you'll read in Chapter 15, the coherence of the decision flows is derived from guidance and boundaries—a strategy.

Decision flows relate to all product line parts, agents, constraints, and forces. Deliberate guidance and heuristics help speed up decision-making. Most importantly, a product line's decision flow must align with and boost the system's other flows.

Contributors and leaders constantly make decisions and choices that affect the line. That's why we view them as a flow. The challenge is that divergent decisions can disrupt all system flows.

Understanding how different players make varied judgments and choices is important. Teams must be able to spot when such decisions diverge.

There's a commonly held belief that decisions should align with an organization's hierarchy. The problem is that this assumption is flawed when those at the top of the hierarchy lack an understanding of the issues at the bottom or when those making judgments about specific projects and system components do not grasp higher-level issues.

The greatest challenge within the decision flow is achieving and maintaining alignment across all judgments and choices. Product line teams must ensure that the other flows correspond with the decision flow and vice versa.

Because some decisions involve many people spread throughout an organization, achieving decision alignment can be challenging. Also, when someone alters any of the processes, whether intentionally or not, the line's management must work to regain alignment.

• Outcome Flow → Strategic Moves

Teams should focus primarily on the outcome flow of their product line. This flow represents the ongoing emergence of factors influencing customer satisfaction, cash flow, and competitiveness. Understand that in product line management, these gains manifest as "Strategic Moves" that add or modify the system's parts.

Our concern involves not just one strategic move but also the development of a continuous stream of moves. We seek strategic initiatives that build on each other and aim to deliver the needed efficiency change in the Accelerating Change Race.

I dedicate Chapters 15 and 16 to creating and carrying out a coherent stream of strategic moves.

Emergence, Flow Resonance, and Shifting Constraints

Outcome flows depend on workflows, information flows, and decision flows.

These flows intertwine and influence one another. There's no guarantee that positive outcomes will emerge or that strategic moves will improve just because the other flows do. However, by strengthening these flows, teams can enhance the likelihood of achieving outstanding results.

Greater gains occur when all flows operate in harmony, creating a resonance that enhances outcome emergence and promotes coherence and synergy across strategic moves. Flow resonance, where each flow supports the others, is a crucial factor in addressing the Accelerating Change Race. It positions change as a viable opportunity within a context that prioritizes efficiency. In Chapter 21, I explore how well-crafted enablers can foster flow resonance.

Flow resonance is a crucial precursor to increasing a product line's Outcome Velocity. This state goes beyond merely aligning the flows. Alignment occurs when the flows do not oppose each other. Resonance happens when the flows amplify each other in ways that would not be achievable otherwise.

The intertwined nature of the flows can make resonance impactful for improving outcome emergence. Consider the impact that improved information flows might have on workflows and decision-making processes. If teams provide better information and insights into the workflows and decision flows, the outcome flow should have a more significant positive impact. I say "should" because a complex system does not guarantee outcome improvement.

The critical mandate in product line systems management is to ensure that flows progress toward enhanced outcome emergence and increased Outcome Velocity. To achieve this, it is essential to understand that if a team's effort to improve their line's outcome flow involves debottlenecking one flow (mitigating a constraint in a flow), another constraint in a different flow will likely surface as the hindrance to achieving a better outcome flow. In complex systems, ongoing mitigation may need to occur across multiple flows simultaneously or in a near-time series.

Flow States: Aligning versus Promoting

Every product manager recognizes the importance of strategic alignment. I don't think I've ever met a strategy or product management expert who can't discuss alignment endlessly. While many managers might roll their eyes at the topic due to frequent lectures, no one disputes its significance.

A significant challenge in product line management is that strategy alignment alone cannot enhance a product line's Outcome Velocity. Product line teams must coordinate their flows to go beyond alignment and achieve improvements across all flows, particularly enhancing a line's Outcome Velocity.

Consider the following "flow states" and their impact on the line's Outcome Velocity.

Flow Chaos refers to the random flow of work, information, decisions, and outcomes. Chaos manifests across the flows, not just within each individual flow. For instance, those involved in projects may perceive the decision flow as disruptive and chaotic. Likewise, those making decisions may consider several of the workflows in the line to be inadequate.

Flow Alignment is the state achieved by intentionally adjusting and matching the product line flows. It implies coherence among the flows. This state requires that a team adopt a shared strategy to establish boundaries that enable coherence across each line's flows. From a complex systems perspective, a strategy serves as an enabling constraint. It facilitates flows by deliberately setting boundary constraints. I share much more about enabling constraints in Chapter 15.

Flow Resonance represents a higher state of alignment where coherency enables each flow to amplify the other flows. Once teams reach this flow state, it becomes quite noticeable.

Each flow state occurs in sequence, building upon the previous one. If team members lack a shared understanding of the status of each contributing flow, it becomes challenging for them to reach the next state. Neglecting flow states also increases the likelihood of reverting backward. Effectively managing flows is essential for overseeing a product line system.

Understanding flow states helps product line teams surpass the strategy of "alignment" preached by innovation and product management experts. The "beyond" states of resonance and promotion reverse the strategy of a business unit. Instead of aligning with the business unit strategy, the bigger issue is for the business unit to keep pace with the product line's outcome emergence. Product lines take a leading role in business unit strategy formation.

System Constraints and Accelerating Change

Constraints impact all complex systems, including product lines. They influence, block, impede, and create obstacles within each flow. They may also establish boundaries and limits to these flows. Unsurprisingly, they can hinder a product line and impede the emergence of the system's outcome.

The major issue in product line management is that rapid change makes understanding flow dynamics significantly more important.

The orderly systems approach involves identifying and addressing the causes of a bottleneck, and once that's achieved, declaring victory. In a complex system, this will not apply to the entire system because there is no single root cause for the impediment. We cannot be certain that alleviating one flow impediment will enhance the overall Outcome Velocity and Change Capacity of the system.

In product line systems management, mitigating flow constraints and maximizing flow resonance work together. However, mitigating a flow constraint may negatively impact the system's flow resonance.

Management of complex product line systems must adopt a different approach, especially under the pressure of accelerated change. By mitigating a constraint in one flow, we need to evaluate if it benefits the line, not just that flow. When the actions prove helpful, teams should encourage more of that behavior. When the actions are not effective, teams should disregard or stop the effort. I refer to this as a weed and feed tactic. It's to "weed out" what doesn't help desirable outcomes to emerge and "fertilize in" what does.

Do Not Toss Orderly System Constraint Mitigation

Teams must also understand that mitigating one constraint often leads to the emergence of new constraints in different flows. Addressing a single constraint may not impact the overall outcome of the line. Instead, achieving a larger outcome may require tackling multiple constraints across various flows.

The challenge in product line management lies in understanding the flows, their alignment, and resonance. It is also important to comprehend the constraints, their types, and behaviors. Plus, teams must be fluent in system enablers, including processes, practices, tools, and environments, which can mitigate or exploit flow constraints to achieve flow resonance.

Chapter 15 explores constraints, and Chapter 21 shares a deeper understanding of product line enablers.

Value Stream Mapping

In the digital world, you'll hear concerns about the development value stream and how teams can speed it up by removing constraints. Understanding it relative to product line flows is important.

The value stream is the step-by-step flow of work to develop or improve a product. It's the mapping that helps managers remove waste from the steps and accelerate the flow. Experts in value stream mapping will tell you it always starts and ends with customers.

Unfortunately, value stream mapping is single-product oriented, disregarding improvements in competitiveness and cross-product gains realized through positioning and platform-leverage. In product line management, insights from value stream mapping would be significant if a team mapped every development project and identified systemic issues across the board. However, I have not seen or heard of a company doing this.

Lean principles support value stream mapping, making the approach indifferent to changing a line's strategy or making strategic moves outside of products. It assumes that product lines will maintain a consistent structure. This focus leads the practice to prioritize near-term, incremental efficiency with minimal consideration for long-term change.

The notion that value streams always begin and end with customers highlights another mismatch with product line management. First, product line systems are continuous; they don't have a definite end. Yes, products will retire, but something else will always take their place. Initiatives often start with technologies and the constant consumer demand for more, cheaper, and faster options. Consider AI. Its value gains have been exponential, but did it originate from a specific customer need? It wasn't until after establishing the technology that companies could tailor it to meet customer needs.

I am not anti-value stream mapping. However, it should not take a primary enabling role in product line systems and their management.

Pace of Innovation

Recently, there has been increased interest in the speed at which companies enhance their products and the processes that support them. This measure is known as the Pace of Innovation. Organizations determine the Pace of Innovation by tallying discrete improvements made over time.

The "Pace of Innovation" is neither a time-to-market metric nor a substitute for product line Outcome Velocity. Instead, it fosters incremental improvement in all facets of innovation and product development. For some, the Pace of Innovation serves as a direct enabler of the outcome's emergence. However, this is not accurate.

Like time-to-market, the Pace of Innovation lacks direction. It assumes that something will improve but does not indicate a purposeful strategic gain. Prioritizing the actions and initiatives that contribute to the Pace of Innovation is an exercise in randomness.

In Chapters 17 and 18, I discuss how product line teams can guide the pace of product line-specific innovation and its effect on the line's flows. This use of the pace of innovation is specific to the interaction among a product line system's agents. It differs from its general application but can make significant contributions to product line systems management.

Flows and The Accelerating Change Race

Every product line has many work, decision, and information flows. All should enhance the flow of the line's outcomes—the emergence of outcomes that contribute to improving the line's Outcome Velocity and overcoming accelerated change

However, conflict between efficiency and change may appear inside the product line flows. For example, some decisions may push toward change, while some work may push toward efficiency.

Every team's job is to spot and communicate how disunity and conflict may happen or are happening within and across flows. They also must create and communicate potential resolutions.

The product line team should seek to gain and maintain flow resonance, where each flow boosts the other flows. They do this using deliberate practices and tools matched to the system's agents and flows.

Chapter 7 Summary

Product Line System Flow

- **Orderly System Practices vs. Change:** Orderly system practices, such as Lean and the Theory of Constraints, may work well in orderly subsystems but can hinder overall change in a complex product line system. Managing product lines as complex dispositional systems is crucial for driving the desired change.

- **Emergence and Flow Resonance:** In product lines, 'flow' isn't a steady stream but rather a turbulent mix that enables outcome emergence. The Outcome Flow depends on three contributing flows.

    1. Workflows include tasks, projects, and initiatives;

    2. Knowledge and Information flows - includes documents, data, and insights;

    3. Decision flows - includes judgments, investments, and choices.

    Achieving coherence and resonance of these flows is powerful. Flow Coherence is the alignment of flows. Flow Resonance creates notable gains when one flow boosts another in otherwise impossible ways. All flows depend on all other flows.

- **Coherent Coordination and Flow:** Coherent coordination within and across workflows, decision flows, and information flows is necessary to achieve flow resonance, boost a line's Outcome Velocity, and ramp up the line's pace of innovation. How teams do this is laid out in the ensuing chapters.

Chapter 8

Common Methods Hold Us Back

Influencing Product Line System Outcomes

We have made remarkable improvements in making and selling products over the past few decades. We have gained a much deeper understanding of how to use technology, develop new ideas, and grasp customer needs. However, we have also created a challenge. Everything we have learned is based on old-school single-product thinking. Now, we realize that this approach is holding us back.

As we learn about product lines, complex systems, and outcome emergence, the limitations of single-product thinking and orderly system methods become clear. Let's explore some practices.

The Single Product Development Blunder

In his 2009 book, The Principles of Product Development Flow,[42] Don Reinertsen outlines a Lean approach to maximizing the flow of product development projects. Reinertsen focused on reducing delays and ensuring the company recouped its investment more quickly. He wasn't concerned with big-picture influences like strategy or generating new ideas. He did not address what happens to products after they launch, except for how much revenue they generate.

Reinertsen's approach appeared very smart. He embraced well-established Lean principles and cost accounting. I remember being so impressed that I spent weeks reading and rereading every chapter of his book. Some experts in systematic cause-and-effect methods still reference the book to demonstrate their knowledge. However, the method never gained widespread acceptance. So, what held it back?

The flow improvement technique Reinertsen offered stood no chance with a complex adaptive system.

His view was on one development project at a time. However, each project is part of a larger system.

Wrong View, Wrong System Substrate

The system is bigger and more complex than Reinertsen assumed. He wasn't dealing with a single-product orderly system, as it seemed he was addressing. Not surprisingly, he confidently declared that companies should employ his causal, orderly systems approach.

The framework and methods were doomed from the start. Complex product line systems will adapt to orderly system practices, and that's exactly what happened. When companies attempted to implement Reinertsen's principles, they discovered that their systems found ways to work around his approach.

The same applies to other practices that don't view the product line as the primary system. A product line is a complex system, which can confuse managers. They tend to rely on traditional, single-product approaches because they don't fully grasp the system. Many companies base their idea generation, screening, and development practices on organized systems thinking because that's all they know.

Screening and Emergence

Screening practices present the same challenge as focusing on speeding up individual projects. They are orderly system practices that cannot thrive within the adaptive development system. In its typical course, a complex system will adapt to avoid an orderly system approach. Go/no-go decision-making and its variants will ultimately seem counterproductive to the complex system. This occurs because the practice interferes with the emergence of desirable outcomes.

It's always interesting to hear experts in an orderly systems approach blame a complex system's adaptive end-run on the organization's lack of discipline or top management's unwillingness to relinquish power. They cite organizational behavior and leadership as the culprits, not the mismatch of their approach to the system.

I have never heard an expert in organizational systems, methods, or techniques say, "My approach fell short because it keeps encountering a complex adaptive system." Why? Most likely, they don't understand the nature of complex systems and cannot recognize when they are dealing with one. I see this frequently in product development and innovation.

Preempting Emergence

Robert Cooper, the creator of the Stage-Gate process and professor emeritus at McMaster University, made the first significant attempt at establishing a "sound" screening tool. This work occurred in the 1980s prior to his introduction of the Stage-Gate process.

Cooper based the tool and screening approach on his seminal "NewProd" study[43], which looked at a statistically significant sample of new product successes and an equally large sample of new product failures. Cooper sought to quantify the impact of over 80 factors identified by academic researchers as influencing a new product's success or failure.

The NewProd screening method never gained traction, despite attempts by companies like Exxon and Procter & Gamble to implement it. (I know because I assisted Professor Cooper with these efforts.) The screening tool's failure does not reflect poorly on Cooper's NewProd research.

No screening practice has withstood the pressures of complex adaptive product innovation and management. Such tools cannot endure because their orderly nature contradicts the system's emergence of outcomes. Companies that incorporate screening practices attempt to declare outcomes without permitting emergence. It's a system end-run waiting to happen.

Screening Versus Emergence

Even now, some three decades later, Cooper still cites executive power and control as the reason NewProd never took hold as a screening tool. I respectfully disagree. The system's adaptive nature provides further insight. Executive behavior and decision-making may influence how the adaptation occurs, but they are not the underlying cause. One way or another, complex product line systems either adapt or die.

We don't argue whether something is a root cause if we assume the system is orderly and not complex. A rationalization that suggests executives who hold too much power and control also hold too much truth to toss it aside. However, a closer examination reveals a different picture. When executives alter the decision flow, they are, in fact, actor agents within the system. They are part of the complex system and contribute to and become influenced by its behavior and adaptive nature.

When we recognize a product line as a complex system, we see issues like top management's decision-making as manifestations of the system's adaptive nature, not as a single root cause. It's like wrestling with a jellyfish. If you squeeze too tight with cause-and-effect methods against an issue, other issues will pop out elsewhere.

Product line management must deal with the adaptive nature of the product line as a complex system. Unfortunately, the root cause analysis tools and methodologies we rely on, all based on orderly systems thinking, are not up to the task.

Actor Agents and System Adaptation

Consider this important insight about system actors. Managers, leaders, and hired experts are actors in multi-product complex systems. In product line management, however, it's best to think of actors as roles and responsibilities tied to skill sets, not as specific people. This helps separate personal relationships from product line management.

System actors can and should influence the system's behavior and emergent outcomes, but they don't control it. While we should expect system actors to influence Outcome Velocity improvements, the complex system will remain adaptive and dispositional, i.e., self-organizing. An actor's influence will prove potent only until the system seeks to adapt its way around the influences. When such influencing becomes ineffectual, you may expect a need for layoffs and firings.

Awareness of product lines as systems and the interplay of their parts and agents is equal to, if not greater than, the importance of leadership styles, work methods, and decision-making techniques. Unfortunately, books, articles, and advisory services related to these other topics don't mention complex system adaptation. You'll even find most academic research to be oblivious to complex system behavior. It is, however, quite real.

Portfolios and Causality

Research I conducted two decades ago showed that the biggest influence on the value of a portfolio of development projects is not prioritization.[44] Nor is it stopping, slowing down, or speeding up projects. Instead, the largest gains come from the projects that enter the portfolio.

The research revealed portfolio management's most significant drawback: It doesn't oversee concept and project generation, which are the most important factors affecting the portfolio's value.

The relation between a project entering a portfolio and the portfolio's value isn't linear. A team of twelve people can just as easily develop an offering that produces $100M per year as they can create a product that delivers just $10M. The difference is the project. It's not the people or the resources.

More Than Development Projects

The nature of new product concepts, the markets they target, and their role in the product line system impact a portfolio's value far more than shuffling resources or reprioritizing projects. Yet, we lose these factors by focusing on improving only the product or the development projects' workflow or expected cash flow.

In today's state-of-the-art portfolio management, projects from different product lines are often grouped. Each project competes with the others for resources, and decisions are made about individual products, not product lines.

Portfolio management pushes strategy decisions to one level above the product lines. As discussed in Chapter 2, this higher level is not "indivisible" in strategy, and it does not set up decisions at the smallest system granularity. That's not what's best for creating and carrying out strategic moves and improving product line velocities. Unfortunately, for all that's been invested in product portfolio management, many managers find it hard to accept that their practices will not yield the best results for their company.

Product Creation, Development, and Management

Companies use many methods and practices to develop and manage products. The goal is to transform technologies into products and services that people desire. To achieve this, teams must ensure that the products or services feature solutions to customers' problems.

Over the years, many insights from smart people have helped us carry out this important work. Two contributions stand out. First, we learned that understanding customer needs is essential before setting attributes and defining technology bundles. Few would dispute this principle. Second, we discovered that orchestrating development work is most effectively achieved through learning and adjusting, rather than rigid task planning. Gantt charts are not useful when outcomes are uncertain. If the outcomes of certain tasks are known, there would be no need to perform them. We can incorporate many other advancements in creativity, resource planning, and governance. There is little doubt that each has enhanced product innovation and management.

But as we move deeper into the twenty-first century and learn to adapt to rapidly accelerating change, not everything is bright. Let me share a few examples of how outdated practices reach their limits and hinder progress.

Processes and Practices

Traditional product innovation and management considered products and projects separate from practices and tools. This natural division seems logical. However, in a product line system, the practices and tools are system enablers and critical system agents. They impact outcome emergence and affect a line's Outcome Velocity.

This understanding suggests a major principle in product line systems management. All agents, including the practices, processes, tools, and methods we use to manage product lines, are as integral to the system as the line's products, markets, and technologies.

The processes, practices, and tools teams must work through also create challenges and opportunities for existing product lines. If the team embraces older practices as part of our product line system, they'll clog up the system's outcome flow. If they adopt new methods that aid and guide the emergence of outcomes, they'll be helping the outcome flow. To build Change Capacity into a line, teams must recast older processes and practices. This job is tough, but it's also necessary.

MBO, KPIs, and OKRs

In the late 1960s, Peter Drucker introduced the world to "Managing by Objectives," or MBO as we knew it. But it took a few decades to find its place inside companies. Everyone had to understand their objectives and how they aligned with other objectives within their organization's hierarchy. Today, that's a problem. We now see that top-level objectives can quickly become nonsensical, making all other subordinate objectives meaningless and problematic. MBO works against rapid change.

Druckerisms also nudged along KPIs and OKRs. The phrase, "What gets measured gets managed," is attributed to Drucker, even though the Drucker Institute tells us he never said that. But why back off from this thought-provoking statement? Too often, it's interpreted as "What can be measured is worth measuring." That, however, we know is nonsense.

The publication of the misstatement encouraged companies to adopt numerous KPIs and OKRs across all business areas, including product innovation and management. Currently, the challenge is that KPIs and OKRs that are effective in an orderly single-product environment can be detrimental in a complex multi-product landscape. For instance, Time-to-Market illustrates this issue, as mentioned in Chapter 4.

When we use methods and processes founded on orderly systems thinking, we confront a tough problem. They don't enable or encourage fluid outcome emergence. Instead, they declare the desired outcome and then tell people to do it. If we continue to use these old-school objectives, goals, and KPIs while trying to keep pace in the Accelerating Change Race, we'll have targeted endpoints that lose sensibility in months, not years. We'll also crash into constraints more often rather than fluidly navigating them.

Examining Our Practices

I've watched many product development practices come and go. QFD (Quality Function Deployment) sparks heated debates among its critics. It originated in the 1970s quality movement led by Japanese firms. It was deeply connected to the Toyota Way and the soon-to-be Lean movement. Product development teams were to use the approach to transform customer needs into product specifications and position requirements against competitive alternatives. It is single-product oriented. Many practitioners are familiar with QFD for its spreadsheet, featuring a triangular "roof" at its top, known as the "House of quality."

QFD led the engineering community to embrace market research in the form of "voice of the customer" (VOC). Those who criticized QFD found themselves in politically incorrect territory. Why? From the single-product, orderly system's perspective, QFD seemed ideal. Its problem, we were told, is that it was burdensome to carry out. No engineering team wanted to invest in sufficient VOC; carrying it out took too much time and effort. That, in turn, spurred many attempts to create "simplified-QFD."

Neither QFD nor simplified-QFD survived. They couldn't. A large company's complex multi-product system always adapts its way around such approaches. This system behavior is fundamental to product lines. Remember this principle as you or your organization seeks to adopt new practices to stay in the accelerated change race.

Addressing The Accelerating Change Race

Traditional product development, which addresses one product at a time, uses orderly systems thinking and focuses only on workflows, and is inadequate for tackling companies' efficiency-change challenges. The issue is not poor use of tools and practices. Rather, it's working with the wrong tools. They don't match a multi-product complex system.

Using complex system tools and practices is foundational to running in the Accelerating Change Race. Such tools and practices are enablers that are part of, not separate from, the product line system.

Chapter 8 Summary

Common Methods Hold Us Back

- **Challenges of Single-Product, Causal-Oriented Management: As** we understand complex system emergence and flow, the constraints of single-product thinking become more apparent. We observe that common product creation, development, and management practices can hinder the emergence and velocity of a product line's outcomes.

- **The Single Product Development Blunder:** Practices focused on speeding development projects no longer work as hoped. Previous generations designed such practices or a single-product orderly system and do not recognize the larger, more complex product line system.

- **Screening Practices and Emergence:** Concept screening aimed at go/no-go decision-making creates problems within an adaptive development system and can be counterproductive to the complex product line system's outcome. The screening approach pushes against the emergence of desirable outcomes by seeking to declare what should emerge before anything does.

- **Processes and Practices:** In Product Line systems management, we must view processes and practices as part of the system, not separate and apart. These are system enablers when they facilitate outcome flow. They are system disablers when they push in the wrong direction. Most old-school single-product processes and practices are system disablers.

- **Portfolios and Causality:** The most significant impact on the value of a portfolio of development projects comes from the projects that enter the portfolio, not prioritization or controlling project speed. The nature of a new product concept and its role in the broader product line system matter most. Common portfolio practices don't embrace the entire product line system, its parts, or possible strategic moves.

- **The Processes and Practices Challenge**: Existing product lines face challenges due to their foundation in single-product, orderly systems thinking. To build Change Capacity, teams must reevaluate their practices and focus on multi-product dispositional systems. Understanding and embracing the entire system is crucial for improving outcome flow and product line Outcome Velocity.

SECTION THREE

Assembling The System

"It may be hard for an egg to turn into a bird: it would be a jolly sight harder for it to learn to fly while remaining an egg."
- C. S. Lewis; British writer, Literary scholar, and Anglican lay theologian

A product line system's assemblage and architecture are as important as its products. - The Author

Chapter 9

The System's Assemblage

Key Parts, Stacks, and Roadmaps

S ome managers, academics, and innovation experts believe a product line is merely the aggregation of many products. Others argue that a product line is a collection of products under a single brand. Unfortunately, neither perspective is correct.

Every product line system consists of many components, not just products or a brand. These components, what the systems world calls "agents," include key parts, enablers, constructors, subsystems, and actors. Some experts argue that flows, forces, constraints, and strategic moves should also be included in discussions about a product line system's assemblage. I fall into that camp.

Efficiency **AND** Change

A fundamental principle in complex business systems is that the interactions among agents dictate the system's performance and its ability to adapt. No single component or agent dictates how the system performs or behaves.

The roll-up of parts does not imply that a system's behavior is just the total of its parts' behaviors. You can't assess the system's Change Capacity by examining each component alone. Instead, the system's behavior is influenced by the many relationships among the parts and the various forces acting on it—some predictable, others not.

You will only understand the system's behavior by observing it in all its aspects—the entire system matters, not just its individual parts.

Many managers struggle to see the full perspective needed to manage a product line system. If contributors and leaders focus only on a single product and an ordered system, they will miss the bigger picture. Product management, engineering, marketing, and business leaders must learn to recognize the components of the product line and how the agent assembly works.

Product Line Systems and its Agents

Today, most companies have only just started shifting from focusing on a single product to considering multiple products. This important shift usually begins when a company implements project portfolio management. However, a bigger step happens when teams unify the roadmaps for all projects and products.

The larger roadmap questions the "whys" of projects. It shifts the focus from project resourcing to understanding and communicating strategic moves. It emphasizes a product line's game plan and strategy by examining the interplay of the system's key components.

The strategy roll-up challenges managers to understand the components of their product line and their interrelationships.[45]

Teams must answer important questions to create a product line roadmap.

- What are the key components of the system, and how do they relate?

- What modifications to the system's structure could enhance the line's Outcome Velocity now and in the future?

- What changes in system assemblage could enhance its ability to execute the line's strategic moves?

Today, we know much more about product line assemblies and their behaviors. They are not mysterious entities to avoid just because they don't fit a company's focus on individual offerings. However, some managers are surprised to learn that system assemblages include more than just products, and that the entire system is just as important as its products and services.

I share a list of primary product line parts in Table 9.1 below. When you include enablers and constructors, a more comprehensive set of product line system components emerges.

Table 9.1: Key Parts Within a Product Line System

Key Part	Description
Noun-based Market Segment	Traditional segmentation schemes like geographies, demographics, and psychographics are taught in business school. Noun-based segmentation is a cornerstone of marketing and selling. A sales team may divide its efforts across geographical Noun-based segments, while marketing groups may focus communications on psychographic Noun-based segments.
Verb-based Market Segment	Clusters of Jobs-to-be-Done outcomes. Verb-based segments are customer-desired outcome groupings that developers deliver product attributes to resolve.

Key Part	Description
Platform-Lever	A common factor that cuts across multiple products in the product line, platform-levers enable leverage (faster development and greater attribute performance bang-for-the-buck). Several types of platform-levers may be used in product line systems. They may be used singularly or in combination with other platform-levers
Product Innovation Charters (PIC)	A common factor that cuts across multiple products in the product line, platform-levers enable leverage (faster development and greater attribute performance bang-for-the-buck). Several types of platform-levers may be used in product line systems. They may be used singularly or in combination with other platform-levers.
Products in Development (PID)	A defined product development project with resources assigned to it, currently under development within the stage/phase-gate and/or agile development process.
Products in the Market (PIM)	A product currently being sold or marketed.
Technology Building Block	A technology that augments or adds value to the platform-lever. Good product line strategies often organize technologies in groups that require similar skills and competencies to advance.

Teams that fail to understand the components, agents, forces, and constraints of their system will struggle to improve the emergence of their line's outcomes. This understanding is foundational for managing product lines as systems. It is crucial when aiming to enhance a line's Outcome Velocity and outpace the competition in the Accelerating Change Race.

New Parts and Strategic Moves

I advise all organizations transitioning to manage their product lines as systems to introduce three key components: Verb-based

Segments, Platform-Levers, and Innovation Charters. Along with several key enablers, such as roadmapping and assemblage mapping, these components enhance the trajectory of Outcome Velocity in ways that would otherwise be impossible.

Table 9.2: Product Line System Agents

System Agent Type	Description / Definition	Example
Key Parts	Components of the product line. Includes: Noun-based market segments, Verb-based market segments, Platform-levers, Products and Services as Products-in-Development (PIDs), Product Innovation Charters (PICs), and Products-in-the-Market (PIMs), and Technology Building Blocks	See Table 9.1
Enablers	An aid to assist or help a system flow (processes, practices, tools, templates, methods)	Product Development Processes: Agile, Stage Gate, Portfolio Management, Design Thinking, JTBD Methods, Key Parts Roadmapping, Assemblage Mapping, Constraint Mapping
Enabling Constraints	An enabler subset. Sets boundaries for flows (work, decisions, information, and outcomes)	Strategy, Innovation Charters

System Agent Type	Description / Definition	Example
Constructors	An enabler subset. Works to create or modify system parts and flow, but is neither a system part nor does it become one. Constructors work to change and improve the system, rather than being part of it.	Tesla's Dojo Computer, LLM Learning Models, PL Engineering
Actors	People, groups, and organizations perform work, make decisions, and use information to aid product line systems management.	Managers, Leaders, Functional Groups, Departments,
Subsystems	These systems operate within the product line system and may be orderly or complex.	Production Systems, Sales Networks, Service Networks
Ecosystems	These are systems that operate outside of the product line system. Chain-link ecosystems are the functional groups within an organization that enable it to operate effectively. A product line is one link within the Chain-link. A Supply Chain is an ecosystem that is external to the organization.	Supply Chain, Internal Functional Chain-link

Managers aiming to boost a product line's Outcome Velocity need to have a strong understanding of competitors and their product systems. Most importantly, they must understand how the parts of their product line system interact to influence outcomes.

Another important characteristic is the assemblage's Change Capacity; a lack of this ability can cause significant challenges for companies when they need to modify their offerings. Often, the absence of Change Capacity is nearly impossible to overcome. This underlying factor is what leads some innovation experts to claim that a core business cannot change.

As the Accelerating Change Race heats up, a system's Change Capacity becomes vital for remaining competitive. I discuss this in Chapter 21 when explaining the Product Line Change Theory and in Appendix 5, where I examine how teams enhance their line's Change Capacity before a change is needed. I keep these topics for later in the book because it's helpful to first learn more about product line systems and their components before addressing these subjects.

The system assembly determines how well a product line performs and how easily it can adapt. In the Accelerating Change Race, both factors are crucial.

The Key Part Stack

You might be familiar with software and hardware stacks. These diagrammatic models are essential for managing software systems, helping developers improve and expand their applications. Stacks also play a similar role in modeling product lines. However, there is one key difference: some elements within the product line stack are conceptual rather than discrete components, unlike in software and hardware stacks.

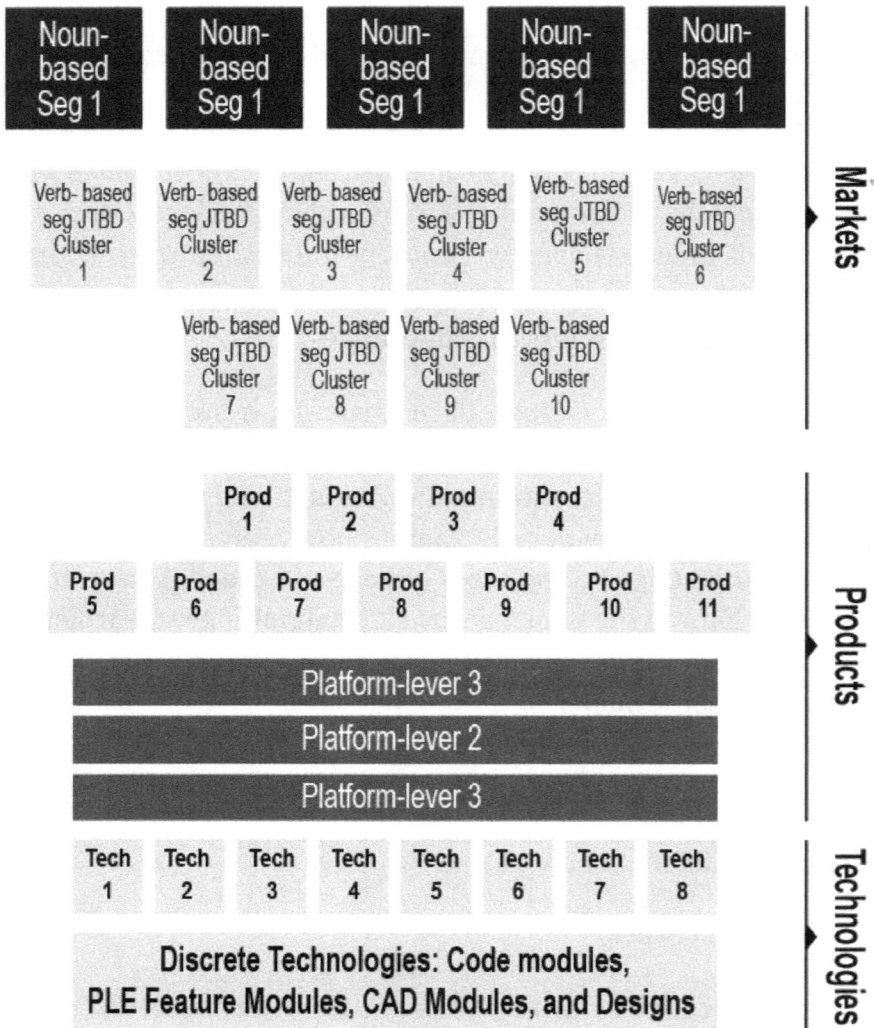

Figure 9.1. Key Parts Stack: *Every product line has a unique stack of key parts. In mature industries, especially commodity industries, competitors often have similar stacks.*

Take a moment to review the stack diagram in Figure 9.1. This visual representation of the stack shows the relationships among the main product line system components. It supports communication and helps create a shared understanding of the system.

From Stack to "Key Parts" Roadmap

The multi-product stack offers a static view of how the team assembles components into a product line. However, adding a timeline turns the stack into a roadmap of key system parts. This visualization is very helpful. See Figure 9.2.

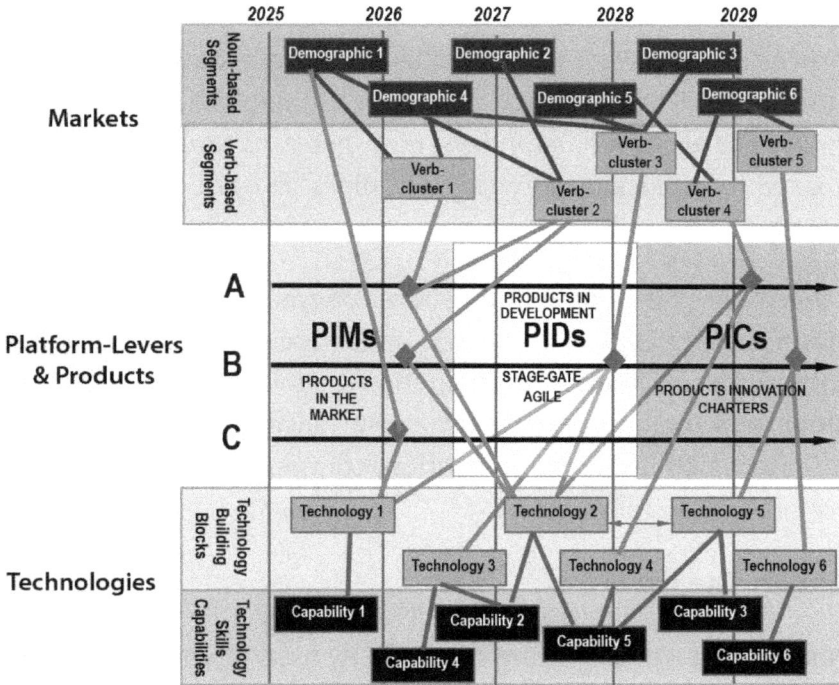

Figure 9.2 0

The roadmap sets the product line stack into motion. Each graphic element on the roadmap represents a key system component. I've added a detailed product line roadmap at the end of this chapter.

A product line roadmap offers a snapshot of the key components of the product line, rather than all system agents. Teams use it to identify and communicate strategic actions. Stacks depict a static situation, while roadmaps show a dynamic one. Note that I am focusing only on the roadmap of key parts. In later chapters, I discuss other visual maps, including Constraint Maps, Enabler Maps, Actor Network Maps, Force Field Maps, and System Assemblage Maps.

Purposeful Communications

Each product line team must decide what their roadmap should communicate. This depends on the roadmap's audience and purpose. For example, suppliers may get a different roadmap compared to those shared with customers or used by teams for planning.

B2B software salespeople often share roadmaps that outline the future of their products. These roadmaps are meant to persuade customers not to switch to competitors, as upcoming software versions will include new features that rivals may promote. If such features do not come to fruition, competitors often call the future product "vaporware."

Roadmap The Parts

Teams develop a product line roadmap by assembling their line's stack and adding a timeline to show planned changes. The chart displays additions, removals, and connections among the components. Usually, the central swim lane shows platform-levers that serve as the timeline for the plot. Visually, this places platform-levers at the center of the product line and highlights the platform-lever as a key factor influencing the system's behavior.

The top swim lane of the roadmap in Figure 9.2 displays the market segments that the teams aim to target with their products. The bottom swim lane emphasizes the technologies that, when integrated into a platform-lever, provide attributes for each product to reach the targeted segments. Each element in the graphic represents a vital part of the system.

Recognize, however, that change impacts all system agents, including the system itself. Some may evolve slowly, but they will evolve. No part of the system remains static, and the list of potential parts is never final. The key components of a product line system you see today will differ in the future.

The parts and agents that companies engage with today differ from those of ten years ago. I'm sure that list will keep changing and look very different in another decade, especially with the rapidly increasing pace of the Accelerating Change Race everyone has entered.

Constraints and Forces

Managing a product line system also requires a deep understanding of key forces and constraints both within and outside the system. These forces and constraints shape how the system is assembled. I introduce this topic here, but I go into much more detail in Chapter 13.

There's an important principle in complex systems that pertains to product lines.

Constraints and forces shape the system. They are critically important in designing and assembling the system's parts and agents. Get it wrong, and the system simply won't survive.

Consider the impact on the product line system when a team conducts Jobs-to-be-Done market research that uncovers a group of underserved outcomes, the Jobs-to-be-Done. They have identified a new Verb-based segment with unmet customer needs, a strong attracting force. Such segments could potentially be a key part of the line's composition. You'll see Verb-based segments at the top of the roadmap in Figure 9.2. However, a new Verb-based segment might require a change in platform-levers, products, and technology building blocks.

Unresolved and underserved needs act like magnets, pulling in their polar opposite products. Customers look for product features that meet their needs. The bigger the customer cluster and the greater the intensity of the need, the stronger the pull on products with the proper attributes. If a team notices this force, they'll likely add a system part (a product or service) and tweak other parts to take advantage of it. Once again, forces and constraints shape the system's assemblage.

Creating Forces

In PLSM, teams create an Innovation Charter, rather than a development project, that guides idea generation based on the compelling research-validated Jobs-to-be-Done. This charter gives purpose and focus to the concept development efforts.

In the complex system's lexicon, the innovation charter is an "enabling constraint." A charter defines the forces and the constraints at play. For example, it may highlight the "Jobs-to-be-Done" attraction within a market segment and the restrictions of a production system or regulatory demand. You can find more about innovation charters and enabling constraints in Chapter 14 and Appendix 7.

Many factors beyond market needs also influence the behavior of product line systems. Refer to Table 9.3. While we wish these influences were stable and followed simple laws, that isn't the reality. Each factor carries a level of uncertainty, and it's often difficult to fully grasp the extent of that uncertainty. Nonetheless, effective product line management requires awareness and understanding of these forces, along with the ability to mitigate, create, and leverage them.

Table 9.3: Key Forces Affecting a Product Line's Assemblage and Behavior.

Forces	Discussion
**Needs, Wants, Desires	Jobs to be Done Outcomes**	These forces attract customers to each product. Customer needs always exist, but the force doesn't come into play until a set of product attributes is available.
Disruptions (slow); Black Swan Events (fast); Weak signal insights.	Other independent forces can disrupt all internal and expected forces.
Regulatory Demands	Regulatory boundaries can shape customer demand and attribute availability. Teams usually know the regulatory demands placed on their products, sometimes based on disruptions.
Risk/Threats (Real & Perceived)	We may view risks as the likelihood of a negative event affecting the system. One problem with risks is when perceptions drive behaviors, causing negative forces to appear.
Superiority, Quality, and Value (Real & Perceived)	The customer's perception of products is a notable force. How well the products match the customer's beliefs about the superiority, quality, and value of each offering reflects a force that either helps or hinders a line's Outcome Velocity.
**Attribute Positioning	Matches, plus Synergies**	Attribute positioning (engineering) and perceptual positioning (marketing) affect customers' attraction to the products. Goal: Engineer a product's attributes specific to Verb-

Forces	Discussion
	based segments (clusters of JTBD outcomes). Target through engineering before communicating positioning through marketing. See Appendix 7 for more about attribute positioning.
Competitive Push, Moves, Blocking	Competitors create forces that work against your offerings in the battle for market share.
Leverage and Scale, Flexibility	Wright's law (the experience curve) states that cost reduction and customer value increase as operations scale, but the ease of change declines unless deliberately addressed.
Organizational Synergy/Alignment	The combined effect of parts and forces is greater than the sum of their separate effects.
Organizational Inertia	An organizational entity will continue its current course until a sufficient force causes it to change.
Resource Capabilities & Capacity: Assets, Money, Skills, and Talent	Having the right resources and the capacity to deploy them into projects can be a positive force toward improved product line Outcome Velocity. Not having them can be a negative force.
Complexity Costs and Overhead	As more products and variants are added to the product line, the total overhead cost per product will increase disproportionately.
Brands (both positive and negative)	Brands and brand loyalty can be powerful forces that affect a product line's connection with customers. Strong competitor brands can be negative forces working against a line.
Diffusion Dynamics	The adoption and acceptance of technology or products by customers and markets. Understanding the forces that induce or hinder diffusion aids in proximate forecasts.

System Constraints and Assemblage

Understanding the forces that influence a product line system is essential. Most importantly, a team's responses to positive and negative forces will likely be ineffective if they don't understand how constraints impact the system's flows.

Complex product line systems encounter many constraints that impact their performance. Some may be deal-breakers, while others hinder flow and emergence.

The assembly of a product line system is directly related to the system's constraints. Think about why this idea is important. The system will fail if the assembly cannot properly handle the constraints it faces. If the system adapts and continues to function, the assembly can be considered adequate, at least for now. The bigger question is whether the assembly will be enough in the future.

A key role of product line systems management is to steer the system toward a more effective future assembly. This vital role involves generating insights to develop and implement changes to the system's configuration.

I'd be remiss if I didn't address the biggest change happening in product line systems and their assembly: the widespread adoption of Artificial Intelligence. It can serve as a product, a platform-lever, and a technological building block—each a vital part of the product line stack—and as a system enabler. Most notably, AI can influence and support the product line as a special enabler called a "Constructor." This new agent type provides system contributions and influences that product line teams have little to no experience in using and managing.

Constructors as System Agents

Constructor Theory is a modern approach to physics that encourages researchers to consider what is possible, what is not, and why. It pushes physicists beyond century-old thinking limited by the laws and principles established by Newton, Planck, Maxwell, and Einstein. It does not disregard these laws and principles; instead, it helps researchers explore what is feasible, discard what is impractical, and then work to understand why. Constructor Theory works backward from reality rather than waiting for new theories to develop.

In product line systems, a constructor is an object or entity that creates what is possible—it makes a process or flow to produce a component that can be added to the system. But it is not required to enable the system to work on a day-to-day basis. For example, Tesla's DOJO computer is a product line system constructor. DOJO is a $1 billion computer customized to process images to create or "construct" a much-desired autonomous driving algorithm (a new platform-lever). DOJO is neither an end product nor a platform-lever, nor is it a subpart of either.

Over the last several decades, product line teams have made notable advancements with platform-levers, their types, and their relationship to business models. Platform-levers have received many multi-billion-dollar investments. With the advent of Generative and Agentic AI coupled with domain-specific knowledge, we're bound to see the same huge investments in product line "constructors."

Undoubtedly, product line teams will leverage AI constructors to develop or enhance system workflows, decision flows, and information flows. The rapid creation of significantly more powerful and faster flows, operating at higher levels of resonance, will fundamentally change product innovation and management. This is not a movement in the Accelerating Change Race that any product line team should ignore. Sadly, it will be a game-ender for those who do not participate. It is worth noting that product line constructors can require substantial capital investments. Currently, we have limited experience managing constructors.

Since AI constructors can create system parts much faster than humans, I believe their use will lead us into a flood of "adjacent possibles." Almost immediately, we'll see new system parts, agents, and products, and not all will be excellent. However, I am confident that constructors will improve and produce much more impactful results. They will be vital for competing in the Accelerating Change Race.

Remember that although constructors may not be the main parts of the system, they serve as essential elements and are connected throughout the entire system. All elements, including constructors, must work together. And to perform well, teams need to oversee AI constructors and how they influence the behavior of the product line system. It's likely that, just as platform-levers are vital in key parts roadmaps, constructors will also play a central role in every product line system assembly.

Assemblage and the Accelerating Change Race

A full product line system assembly includes all agents— parts, actors, enablers, and constructors. The full set of agents dictates the system's behavior, rather than just the sum of each agent's behavior. This implies that all agents must be considered as integral to the system, rather than separate entities. Understanding the full system is crucial as a team builds out the system's assembly and strives to implement impactful strategic moves.

Keeping pace in the Accelerating Change Race takes a full systems perspective, not just a focus on key parts. This view includes the key part stack, system enablers, actors, and constructors.

Figure 9.3 on the full page following this chapter's summary displays a full product line roadmap, less company-confidential information.

Chapter 9 Summary

The System Assemblage

- **Complex Systems:** Product lines are complex systems with many interconnected parts. These include products, core components, enablers, constructors, subsystems, and actors. How these parts interact affects the system's efficiency and ability to adapt.

- **Multi-Product Thinking:** Understanding a product line system means thinking about multiple products instead of just one. This involves planning and knowing how all parts of the system work together, which cannot be understood by looking at each part separately.

- **Key Parts Assemblage-The Product Line Stack:** Key parts of a product line system include Noun-based market segments Verb-based market segments, platform-levers, products-in-development, products-in-the-market, innovation charters, and technology building blocks. The way these parts relate and assemble is called the product line stack.

- **Product line roadmaps:** These graphics show how the system key parts change. This roadmap helps teams plan and communicate strategies. They are tailored for different audiences, such as suppliers, customers, or internal stakeholders.

- **Forces and Constraints:** These critical influences shape a product line system assemblage. Managers must understand these forces to improve the system's efficiency and adaptability.

- **Constructors:** These are special system enablers that can make or construct system parts, agents, and flows. They have become especially relevant to product lines because of AI. An example is Tesla's DOJO computer. Tesla designed it to create an autonomous driving platform-lever, a task that would be much more difficult and time-consuming without the DOJO constructor.

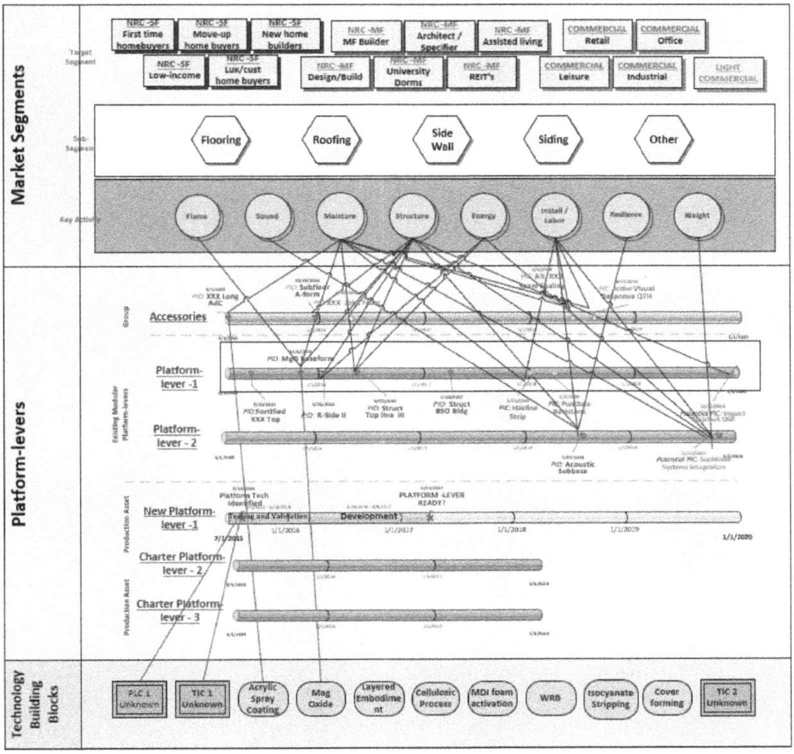

Figure 9.3. Full Product Line Roadmap - *Please refer to the online appendix for a full view of this product line roadmap.*

Chapter 10

System Architectures & Behavior Patterns

Structuring for Efficiency and Change

The architecture of a product line system refers to the specific form or structure of its assemblage. It is a key determinant of a product line's behavior, performance, and capacity for change.

Here's a key insight:

> *Because the system's structure creates boundaries that determine how a product line behaves, you must change the system's architecture if you want significantly different outcomes. When seeking such change, teams must have a deep understanding of their line's current and possible future architectures—the structure of their line's key parts and agents.*

In Chapter 9, I discussed assembling the components of a product line system. Here, I share how the structure of a line's parts and agents influences its performance. The next chapter investigates different product line system architectural types and their impact on strategy, business models, and organizational structures. Understanding the system's architecture and the behavior that emerges from it is crucial for competing in the Accelerating Change Race.

Architectural Change

Every product line has a system architecture. However, most managers don't prioritize this system design until the product line underperforms or, as we say in product line systems management, until the line's Outcome Velocity drops. At that point, many teams realize their product lines are not easily adaptable or changeable. The most common problem is that the capacity for change was engineered out of their product line system's architecture to achieve greater efficiency.

A product line's Change Capacity is a feature of its architecture. High Change Capacity lowers the cost and time required to modify the system, its components, and agents, as well as their interactions. The challenge is to identify the need for change in the system's design before a line's survival depends on it.

A product line's architecture is more than just a stack of key parts; it also includes actors, enablers, and external subsystems. These agents not only enable the system to operate but also act as the connectors that hold the system together.

As I explained in Chapter 9, there are several types of agents in a product line system. Actors are individuals or groups of individuals who contribute to the product line. Enablers are assets, processes, methods, and IT supports that aid a product line's flows. Constructors are a specific type of enabler that helps "construct" system parts and recreates system flows by carrying out work, transferring information, and offering reasoning in decision-making—all in non-human ways. It is up to teams and their companies to decide how to assemble the parts and agents. We refer to this deliberate structure as the system's "architecture."

Table 10.1 lists the potential parts, actors, and enablers of a product line.

Table 10.1 Product Line System Parts, Actors, Enablers, and Aids

Key Parts	Actors	Flow Enablers	Practices, Tools, Aids
Noun-based Segments	Part Teams	Responsive Roadmapping	Lean
Verb-based Segments	Suppliers	Roadmap	Theory of Constraints
Platform-Levers	Advisors | Topic Experts	Charter Formation	Design Thinking
Products	Functions | Chain-link	Insight Affordance	JTBD
Product Innovation Charters	Ecosystem Contributors	Org Structure	Delphi Casting
Products in Development	Suppliers	Performance Reviews	Backcasting
Products in the Market	Distributors	Agile | Scrum | Backlogs	Custom Algorithms | Data Analytics
Platform-Lever Charters	Service Firms	Portfolio Management	Machine Learning | Artificial Intelligence

Key Parts	Actors	Flow Enablers	Practices, Tools, Aids
PLE Feature sets (library)	Business Unit Leaders	Waterfall | Stage & Gate	System Deconstruction
Technology Blocks	PL Governance	Concept Generation	UX
Technology Modules	PL Orchestrator Team	Distributed Leadership	VOC
Functions | Chain-links	Partners	DevOp	Technology Readiness
Ecosystem Contributors	Developers | Engineers	IT System Supports	Life Cycle Analysis
Constructors	Product Managers	Project Management	Outcome Velocity Measures & Reports
	Project Managers	Process Management	Financial & Cost Accounting Models
	PL Leaders	Knowledge Management	Customer Satisfaction Research
	Constructor	ERP Integration	Competitive Analyses
		Dashboards	Attribute Planning & Targeting
		Multi-Generational Product Planning	Knowledge, Info Repository
		Business Model	Charter Templates

Key Parts	Actors	Flow Enablers	Practices, Tools, Aids
		Skills & Capabilities	Dashboard | Reporting
		Constructor	

Visualizing Architecture with Actors and Enablers

A pictorial diagram of a line's architecture can be quite large. If placed on one sheet of paper, it's likely to appear as an "eye chart." A top-level view shows the key parts stack, along with key actors and enablers. See Figure 10.2, which shares a top-level, rudimentary view. A more detailed view would be displayed as a network diagram showing the relationships and interactions among all the system parts and agents.

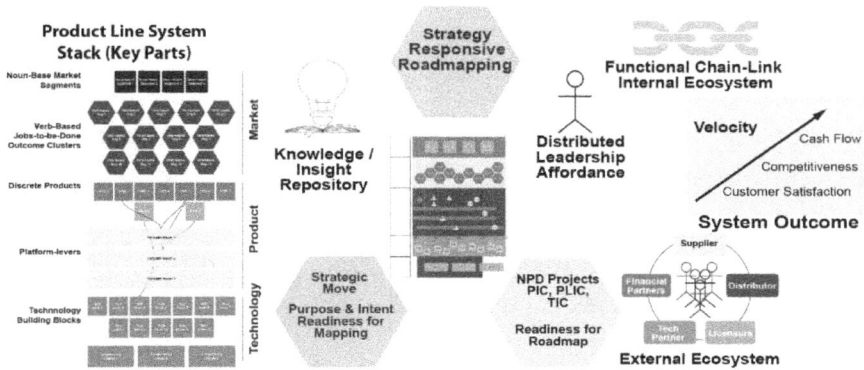

Figure 10.1. *The Product Line Architecture encompasses the stack, actors, and enablers. A large view of this infographic may be found in the online Appendix.*

Roadmaps and Architecture Layouts

As discussed in the last chapter, the product line stack acts as a foundation for creating the roadmap. The architecture is a schematic representation of the product line and the product line stack. It does not include a timeline. It also shows actors and

139

enablers to clarify how the system functions. Like product line roadmaps, there are no specific graphic standards for presenting the layout.

Teams may also want to create network diagrams that illustrate how the components and agents are connected. You'll discover that creating these diagrams can provide valuable insights into the system.

The product line architecture and its management are only meaningful when teams have a means (an enabler) to influence and guide the improvement of outcome emergence and the line's Outcome Velocity. I call this enabler System Assemblage Mapping. It's an ongoing process that merges product line roadmapping (the stack and key parts) with modifications to the architecture and flows. It helps teams enhance the entire system, not just its key parts.

System Assemblage Mapping is a process that enables other enablers. It promotes integrated coherence and influence across the entire product line system. I mention it here because of its importance in advancing a product line's assemblage and behavior. I discuss this process in Chapter 22 and Appendix 7, and I provide example assemblage maps in the online appendix.

Deconstructing the Architecture: Exposing Behaviors

One key principle in PLSM is regularly disassembling and reassembling a product line's architectural model. This exercise focuses on conceptual models of assembly, not the actual system. The task involves breaking down the system's actors, enablers, constructors, and key components, and analyzing how they interact with each other and with the system's forces and constraints. It's about identifying what fits and works or helps and what doesn't.

An analysis of the entire system can reveal many aspects of its behavior, including what, why, and how. The task requires a thorough understanding of the product line and its agents. Those conducting this important work must identify their product line's drivers and limiters—the system enablers, builders, and constraints—that affect the line's performance and ability to change.

The difficulty in doing the work lies in our frequent misjudgment of the forces and constraints, as well as in our misjudgment of the need to retire old enablers and develop new ones. These misjudgments come from our assumptions or a lack of clear understanding of the forces and constraints involved. An enabler may be labeled as a "best practice," but that "best" status might no longer apply as forces and constraints change. A team's sensitivity to enablers depends as much on their knowledge and insights into forces and constraints as it does on their assumptions and beliefs about what's best.

Designing for Change

A line's architecture is essential for adapting to rapid change. Ideally, the system design allows for change while maintaining efficiency. One way to improve a line's ability to adapt is to incorporate that capacity into the system's architecture.

A line's Change Capacity comes from the inner workings of the product line system—the system's modularity, flexibility, and adaptability. However, Change Capacity is not a specific part of the system. Instead, product lines gain Change Capacity through the dynamic interaction of system components and their connections with enablers, constructors, and actors.

The extent to which change is possible and easy to implement may seem like a trade-off with the system's efficiency. But that's not the case. Limitations on change arise due to technical or organizational constraints among the system's agents and components. You might blame efficiency, but the real cause lies in the system's agents, its key parts, and how everything interacts.

The major challenge for many lines is that the standardization imposed on their systems to increase efficiency also causes parts and agents to become rigid. Platform-levers are the most common system components that get locked in place due to standardization in the pursuit of efficiency.

Consider a company with a large manufacturing site that produces various products using a common engineering design—a design platform-lever. Production standardization is specific to the design platform-lever. This work intentionally enhances efficiency. However, when that is the only action taken, changes to parts and agents may reduce the line's ability to adapt. This outcome is unintended. Still, it results from our traditional thinking, which prioritizes efficiency above all else.

In complex systems, the system's behavior establishes constraints and boundaries regarding when and how a product line system can change. These restrictions may hinder a team's ability to respond effectively to accelerated change.

Product Line Engineering

Many engineering groups adopt a powerful approach to improve flexibility and Change Capacity in their product lines. It's called Product Line Engineering. This method acts as a system constructor that helps engineering teams directly add to the line using predefined modular blocks. I refer to this feature of a line's design as "composability." It uses software to allow engineers to quickly incorporate pre-designed modules into a core product design.

Those who use PLE refer to the add-on building blocks as "feature modules." A complete set of these modules is called the 'feature library.' These modules resemble technology building blocks in the assembly line.

Several software vendors offer tools and methods that support PLE. These tools help in creating tangible products (CAD/CAM designs) and ensuring software compatibility. PLE acts as a system constructor rather than just an enabler, because it actively constructs a key component (products) directly.

Design engineer Charles Krueger deserves much credit for developing and refining PLE[46]. Today, PLE is recognized by the International Council on Systems Engineering, the International Organization for Standardization (ISO 26580:2021), and the International Electrotechnical Commission.

AI-Generated Compositions Blocks

PLE is a powerful system constructor that supports hardware and mechanical product lines. Similar constructors exist for software, such as code modules and APIs. The blocks can be physical or digital, including software-tested digital twins for hardware and sandbox-tested software components. When we incorporate such composite blocks into the product line stack, they sit between technologies and products.

Ready-to-use composition blocks are effective only if they complement a platform-lever and enable products to meet customer needs. These blocks are common in product lines that offer customization or engineered-to-order products. This suggests that a service-type platform-lever may be necessary in the system. Refer to Appendix 6 for a more detailed explanation of why this is the case.

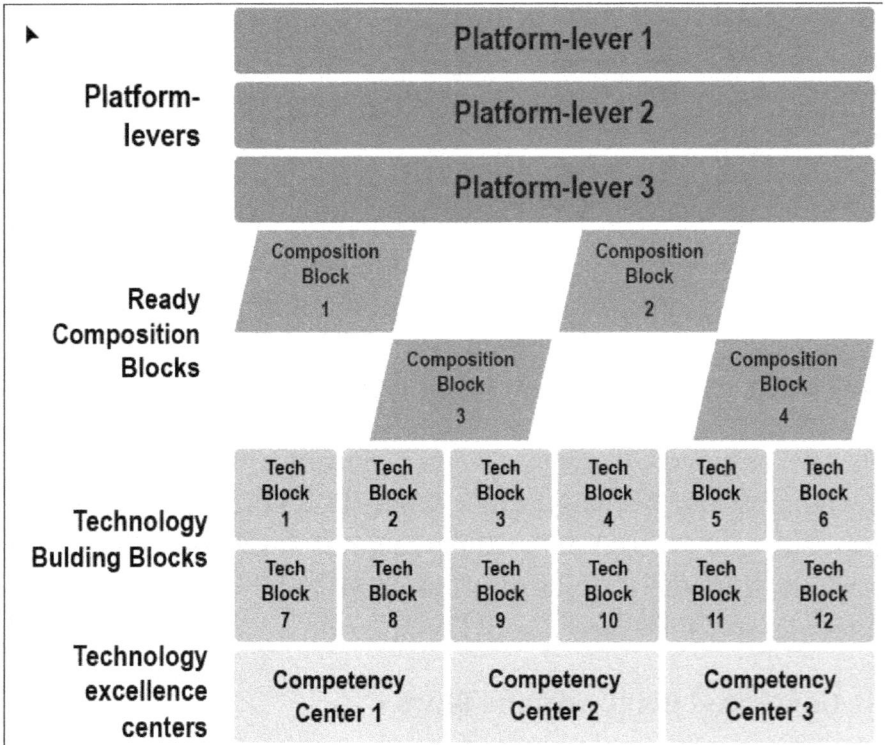

Platform-levers	Platform-lever 1					
	Platform-lever 2					
	Platform-lever 3					
Ready Composition Blocks	Composition Block 1		Composition Block 2			
		Composition Block 3			Composition Block 4	
Technology Bulding Blocks	Tech Block 1	Tech Block 2	Tech Block 3	Tech Block 4	Tech Block 5	Tech Block 6
	Tech Block 7	Tech Block 8	Tech Block 9	Tech Block 10	Tech Block 11	Tech Block 12
Technology excellence centers	Competency Center 1		Competency Center 2		Competency Center 3	

*Figure 10.2. 3*Generative AI can create text, images, software, and CAD/CAM models. By combining machine learning with domain-specific knowledge bases—such as Retrieval-Augmented Generation (RAG) for industries, markets, technologies, and product lines—product line teams can also develop composition blocks and run tests on each.

Simulation tests, in particular, can help teams determine how effectively the composition block integrates with the product line system. Teams must then address a critical question: how does the composition block help align a product with a customer's needs, and how does that impact the product line Outcome Velocity vector?

Building Change Capacity

Most teams prefer not to have to battle their system when implementing change. They also want to avoid dealing with late strategic moves because their system adapts poorly. To prevent these issues, they must create an architecture that supports change.

We recognize that customer needs and technologies will now evolve faster than ever. The product line team's role is to anticipate these changes by understanding what, where, and how before things happen. Futurists excel in this role, leveraging every available tool to create possible scenarios.

But the uncertainty of our complex product line system will continue. We won't be able to perfectly predict the future of markets, industries, and technologies. Therefore, our approach should use futurecasting[47] techniques and develop sensitivity to potential changes, observe what changes do happen, and then respond accordingly. In the new world of rapid change, locking in on actions too early can be as harmful as acting too late.

Alacrity and Change Capacity

Product line teams must consider the question: "Can we change any system part or agent so it more easily adapts to future influences?" Think about Apple's M series chips. This platform-lever chip works with macOS in Apple's computer line. Apple engineers designed a processing component to handle machine learning and AI within the chip. This is CPU processing that supports essential ML and AI functions without wasting time or power calculating hundreds of decimal points.

Few customers used the ML and AI features when Apple launched the first chip (M1) in late 2020. This important design also paved the way for the M2, M3, and M4 chips, which provide improved machine learning (ML) and artificial intelligence (AI) capabilities. Such functionality is likely to continue expanding. Also, note that Apple also makes changes to macOS that align with the M chip updates.

Building Change Capacity involves more than just a technical challenge; it also presents issues in product line architecture. Managers need to evaluate whether current and future agents and components will support or hinder a line's performance and its ability to adapt.

Architecture and The Accelerating Change Race

All teams will find it challenging to address the need for simultaneous efficiency and change. To achieve this, they must thoroughly understand their system's architecture. This structure is a critical factor in a line's efficiency and capacity for change. It determines both the system's Outcome Velocity and its Change Capacity.

Without the capacity to change, a line will stagnate, and the business will not realize its full value.

Chapter 10 Summary

System Architecture and Behavior Patterns

- **Product Line System Architecture:** This conceptual model explains the assemblage of a product line system.

- **The Accelerating Change Race:** The product line's architecture is central to overcoming efficiency and change simultaneously. It enables a line's performance and its capacity to change.

- **Stacks, Architectures, and Roadmaps:** Every product line has an architecture. Teams depict their line's architecture as a stack that shows market segments, products, and technology blocks. The architecture includes actors and enablers influencing the system. Roadmaps complement architectures, revealing a product line's dynamics and intentions for advancement.

- **System Assembly Mapping:** This important practice enables teams to orchestrate the change of their product line architecture. It purposely helps teams address both efficiency and change by modifying the stack, plotting the key parts roadmap, and transitioning system enablers and actors.

- **Designing for Composability:** A crucial aspect of a line's architecture is its composability, which allows the easy composition of new offerings. Composability extends from platform-levers and technology blocks to data structures, materials, and regulatory demands.

- **Architecture and Flows:** The architecture affects the product line system's flows — cash, information, work, decisions, and outcomes. Teams must assess how the architecture improves or impedes these flows and whether it promotes flow coherence and enables flow resonance.

- **Building Change Capacity:** Creating an architecture conducive to change involves considering future scenarios, multi-generational platform-lever planning, and addressing agent "anti-patterns" that have the potential to hinder change.

Chapter 11

System Archetypes, Business Models, & Strategy

Matching Product Line Archetypes and Business Models

A product line's architecture greatly influences how the system's components and agents interact. It is a key determinant of the system's behavior. The connection between architecture and behavior underscores an essential aspect of product line systems.

When product line system architectures are similar, their behavior patterns also tend to be similar.

This insight highlights an important characteristic of product line systems. Because we need to manage a line based on its behavior, the system architecture acts as a guide for how to best manage the product line and how the company's functional groups should interact with it.

By classifying or grouping product line architectures based on key system components, we can determine the overall management approach that best suits the product line system. We refer to these structure-based classifications as "system archetypes."[48]

In complex systems theory, the archetype acts as the core structure of a system. The system develops and adapts from this initial base structure.

Platform-levers and System Archetypes

Because a product line's archetype greatly influences the system's behavior, it can be the most important factor that product line teams and their organizations must address to improve a line's efficiency and adaptability.

The core of a product line stack, the line's platform-lever(s), is a key determinant of the system's archetype. In strategic planning, when a team chooses to focus on a specific bundle of technology, assets, or services as a platform-lever, it is also selecting the product line system's archetype. With that in mind, let's examine some of the platform-lever types that teams might consider.

In the last chapter, I discussed how every line should have at least one platform-lever. Without this essential component, the line will be weak and ineffective. It is central to the product line's strategy, stack, and roadmap. Table 11.1 lists several types of platform-levers. You'll also find Appendix 6 dedicated to platform-levers.

Table 11.1 List of Major Platform-Lever Types

Platform-Lever Type	Platform-Lever Example	Leverage Source
Production Asset	A chemical reactor, a paper-making machine, and just-in-time automation	Scale or flexibility
Hardware Design	A computer motherboard, a system's core controller	Design reuse
Service System	An automated bank teller, an automated car wash	Automation, speed
Software Systems	A software operating system, an Integrated application software	Versatility within the intended domain
Proprietary Formula	A unique pharmaceutical, molecular structure, formula, complex system	Uniqueness
Embedded Infrastructure	An optical fiber network, a social network	Connectivity
Modular Systems	A roofing system, a closet connection system, and integrated automation equipment	Adaptability in use, multiple uses
Algorithms, Generative AI, AI Agents, and Machine Learning	Bank loan screening models, search engines, and voice recognition	Rules, judgment, analysis, decision acceleration, understanding
Connected Integration (IoT)	Intelligent and integrated HVAC (heating, ventilation, and air-conditioning) controls	The data structure, stack, and queues
Blockchain	An NFT that assigns certification to a tangible or intangible product or object.	Scale and meta transfer
Hybrid and Combination	An automotive assembly (with engine and frames as separate platform-levers)	Multiple

Consider how a manufacturing process, an operating system, and a hardware design are three of many types of platform-levers. Each differs from the other two. And, for each, the product line team must establish other key parts, enablers, and governing constraints to form the system. They must also facilitate the integration and interaction of everything. The platform-lever, however, will dictate much of the system's behavior.

Each platform-lever, regardless of its type, is a deliberate choice that is a cornerstone to a product line's strategy. Don't view them as naturally occurring phenomena that last forever; it's quite the opposite. To fulfill their purpose, organizations must invest in them intentionally and focus on them intensely.

A functioning product line system can have multiple platform-levers. However, since each requires intense focus, having more platform-levers within a line makes it increasingly difficult to focus on each. Also, each platform-lever has a lifecycle that, as it matures, can change the system's behavior. As AI-driven change accelerates, the lifecycles of all system parts, including platform-levers, will continue to shrink, demanding more attention to the interactions among system components and agents.

Apple's Platform-levers

Let's explore platform-levers using Apple's product lines, which I introduced in the last chapter.

More than a decade ago, Apple established internal chip design capabilities they called Apple Silicon. Its purpose was to design CPU chips tailored to their respective product lines. The platform-lever they targeted was a combination of CPU chip and an operating system. For the company's computer line, you would recognize it as the "M" Chip-macOS combination.

Apple's A-series chip and iOS form the iPhone's platform-lever. Everything else in each iPhone is built on this combined platform-lever. Notice how Apple continues to enhance the unique chip and operating system combination across each of its product lines—iPhone, Computers, AirPods, Watch, and VisionPro. Since each platform-lever type is the same across all lines, this deliberate design implies that these lines also have the same system archetype.

We also see Apple preparing to change this chip-operating system platform-lever mix—each product line's archetype—to include Apple Intelligence, reflecting the company's user-focused, customer-centric approach to AI.

Apple Intelligence is designed to work seamlessly with its chips, as each has a dedicated processing area optimized for AI and Apple Intelligence. As Apple's platform-lever combinations develop, we can expect the company to introduce the A-series chip, iOS, and Apple Intelligence as the new platform-lever mix for the iPhone.

Please keep in mind that a platform-lever is just one key part, though a very important one, in a product line system.

Platform-levers Define Archetypes

We identify a product line's archetype by recognizing its platform-lever types. And, like at Apple, a product line can include multiple platform-levers, each potentially different.

Consider this for a moment. Because the behavior pattern of a product line system is heavily influenced by the specific combination of the line's platform-lever types, we can identify a product line system archetype by examining the system's platform-lever types, whether there is just one or multiple. As we will see, this understanding is crucial for managing both efficiency and change within a product line.

Archetypes and Strategic Moves

Understanding product line system archetypes is essential for advancing a line. It also helps in planning your strategic moves against competitors' likely actions.

When considering strategic moves, it's clear that everyone wants new technologies to ramp up quickly and boost a line's Outcome Velocity. However, there is no direct link between technology and the emergence of outcomes. Outcome emergence — the desired improvement of the line's Outcome Velocity — results from the interactions among all system components and agents.

Technological opportunities can raise many system assemblage and architecture questions for product line teams. Should the technology be allowed to mature and serve as a building block that enhances an existing platform-lever, or should it be integrated into a new, archetype-changing platform-lever? How will the new technology, combined with other system parts and agents, impact customer satisfaction regarding their Jobs-to-be-Done? What competitive responses can be expected? How might it impact cash flow? The answers to these questions depend on the product line system, its archetype, and the energy, investment, and time required to implement the necessary system changes. Plus, as we'll see, it is also greatly dependent on the business model in which it must operate.

Product Line Architecture and Business Models Congruency

In complex systems, such as a product line, the system architecture is the primary influence on the system's behavior. Similarly, a business model acts as the architectural determinant of the organization's operational interactions, that is, its behavior. For a product line system to perform effectively, its behavior must operate in harmony with the behavior of the organization. This means the product line archetype and the business model type should be aligned.

The business model, specifically the type of business model[49], defines the relationships among the organization's functional groups, with the product line being one of these groups. However, product lines are more like the queen bee of functional groups. A product line's performance is crucial to a business unit's performance. This makes the relationship between a product line archetype and the business model extremely important.

Let's take a deeper look at the relationship between a product line's archetype and the business model.

Product Line Congruency Theory

Ongoing operations require teams ensure their product line behavior aligns with the business unit's behavior. In keeping with such alignment, I proffer The Product Line Congruency Theory, which states:

> *A product line's archetype and an organization's business model must operate in congruence for the product line system and business to operate efficiently. Consequently, any change to one requires a corresponding change in the other.*

For those who require a more concise understanding of the key connection between product line system archetypes and a business model, you'll find a deeper dive in the online appendix. Those in the software world may notice how it somewhat (but not completely) reflects Conway's Law[50], while those with an academic background may see a reflection of the Mirror Hypothesis[51].

As I discussed, the key determinant of a line's archetype is its platform-lever type(s), and the "how" of a product line strategy will always seek intense focus on the exploitation of the platform-lever(s).

If a team or organization requires different outcomes from its product line, it must adjust the product line strategy and system architecture. When such changes include a change in platform-lever type(s), the system archetype will change, and thus, the business model will require adjustments. See Figure 11.1.

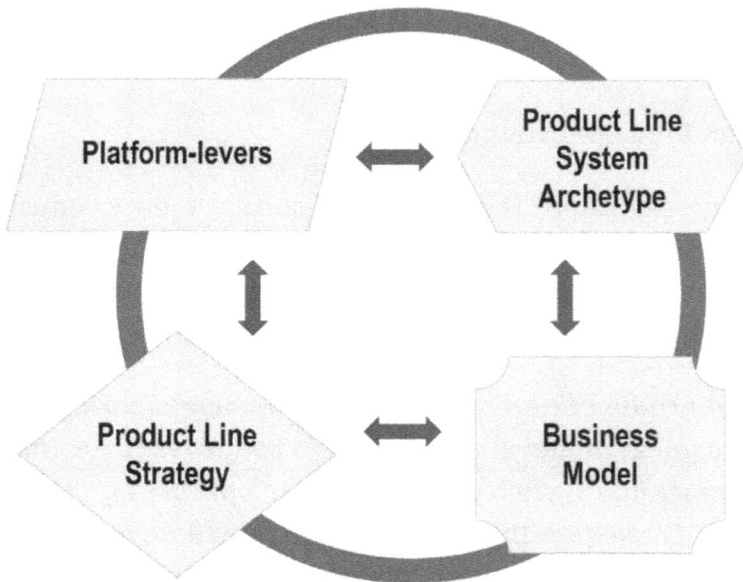

Figure 11.1 *High performance demands alignment of the product line archetype, platform-levers, product line strategy, and business model. Figuring this out is an iterative process.*

Often, a product line may incorporate two or three platform-lever types, creating a hybrid or mixed system archetype. For example, this may be a manufacturing platform-lever combined with a service platform-lever, or standardized production running in parallel with design-to-order manufacturing. The Product Line Congruency Theory then suggests these mixed or hybrid product line archetypes must have corresponding dual or mixed business models.

However, problems can arise when more than one product line system, whether of the same archetype or not, must interface with one business model, whether singular or dual. The issue is that the functional interplay guided by the business model will become strained in a tug-of-war between the product lines. This puts one or both product lines at a competitive disadvantage.

Perhaps the most concerning aspect of our Accelerating Change Race is that product line system archetype changes will become increasingly commonplace. Yet, the product innovation and management communities remain single-product-oriented, unaware of product line systems and their archetypes, and the business model community is blind to product line systems and their archetypes. That, quite obviously, must change.

Determining An Innovation's Fit

A corollary to the Product Line Congruency Theory can be particularly helpful, especially for teams managing corporate innovation initiatives.

We can foretell how well a major innovation will fit into a business by comparing that innovation's system archetype to the archetype of a line that is operating successfully within the business model.

This statement includes key insights about innovations. It indicates that product line archetypes can help us better understand why innovations do or don't, can and can't, become part of businesses. It also suggests that the product line archetype is critical to how we plan, organize, and structure innovations for success.

A key challenge for introducing a major innovation into an existing line is that if the innovation is product line archetype-changing, it may become "blocked" by existing business models and the organizational behavior induced by those models.

Major Product Line Changes

Archetype changes in large companies are challenging undertakings. While reading a business article or attending a conference presentation might make these strategic shifts seem straightforward, they are not. Such changes involve significant technology, marketing, and organizational hurdles that demand that managers stay clear-headed and show considerable courage.

An archetype change also presents a significant leadership challenge. Many leadership experts and coaches offer guidance on navigating this change, but only from the crucial human-to-human perspective. However, from a systems point of view, it's important to recognize that leaders are agent-actors. They are part of, not separate from, the system.

Leaders must engage with new agents and enablers, sometimes temporarily and other times more permanently. They must facilitate the system's new flows, components, and emergent outcomes. The adaptation of various system components and agent behaviors is vital when developing a new system assemblage. Leadership is just one component. I explore this in greater depth in Chapter 22.

The Continuously Changing Archetype

Consider Tesla in 2024, before Elon Musk entered the U.S. political scene. At that time, Tesla remained well ahead of the major players in the automotive industry. For many years, the company tirelessly worked to expand its market lead by intentionally changing its product line system, not just by improving the efficiency of its products. Change and efficiency went hand-in-hand.

From an outside observer's perspective, Tesla transformed its EV product line by radically redefining its product line archetype, a move enabled by changes to its platform-lever mix. The company first merged its battery operations and efficient motor into a single drivetrain platform-lever, as shown by their "Gigafactory" in Nevada, with Capex estimated to surpass $4 billion. The battery gigafactory is a joint venture with Panasonic and began full operations in 2016. This unified platform-lever defined their system archetype, allowing industry disruption.

Next, the auto company developed its new system archetype based on three platform-levers. The first was the battery-motor drivetrain. The second platform-lever was the integration of an electronic transmission and full-car electrical controls. The third platform-lever was "Gigacasting," a manufacturing process in which they cast the car's underbody as a single part, rather than bolting together multiple parts. This new combination of platform-levers established a new product line archetype.

Currently, Tesla is launching and refining a new software platform-lever that leverages an autonomous driving algorithm, once again redefining its product line archetype. This will also necessitate changes to many system agents and enablers, as well as to the company's business model and product line strategy. We also hear murmurs of an automotive industry that may rent car features, for short periods, to match each customer's changing needs. That, no doubt, is a much different archetype that demands a much different business model. Do you think the old-school auto companies can keep up?

System Architecture and Value Capture

Many B2B companies face a major systems challenge with novel innovations, especially in the chemical, materials, and software industries. It happens whenever a business tries to connect its new, novel services or products with those of another company.

The challenge for the innovating company is to figure out how to capture value from the final product—the customer's product. The solution isn't simply replacing an old part of the customer's offering with a new one; that suggests the customer would only buy it if it reduced their costs. To gain more value, the customer needs to redesign their end product to take advantage of the new product and create value in the offering for the end customer— the customer's customer.

The line's innovation will fail if customers do not redesign their products.

Somehow, product lines and sales teams must persuade customers to redesign their end products to incorporate the new product. If customers struggle with this task, the innovative product won't quickly penetrate the market. However, there is a solution. It involves adjusting the new product's value proposition and the line's archetype. The offering must be: "We provide business customers with a high-value product and a design service to help our customer capture the value of our innovation." Achieving this requires reworking the line's architecture (a change of archetype) to add a design service platform-lever. This strategic move is essential.

Major Archetype Moves

Jack Welch adopted this product-plus-service strategy when he took charge of GE Plastics. The company created lighter and stronger plastic, but convincing customers to buy it was difficult. While using the new plastic would make the customer's end products lighter, smaller, and stronger, most potential customers faced a tough choice.

The issue was that GE's customers had to redesign their end products if they wanted to benefit from GE's new plastic. For example, the new lightweight material could have allowed customers to modify a product's housing or internal parts, enabling them to make a motor smaller or use less power. To effectively enter the market, GE needed to empower its customers to redesign their products. Therefore, the GE plastics team paired the polymer production platform-lever with a service platform-lever dedicated to helping customers redesign their end products. This strategic shift changed the typical design of GE's product line system, and its success helped advance Welch's career.

System archetype changes across all industries are more common than you might think. For example, digitalization appears to be in a constant state of flux. Nearly all software companies have transitioned their offerings to cloud-based solutions, and the value delivered to customers has been extraordinary. The shift to a cloud-based platform-lever move was a fundamental change to the archetype. However, it's important to note that each customer's practices also had to change if they wanted to gain value from the new cloud platform-lever and archetype change.

Archetype Behavior Patterns and Anti-Patterns

The behavior and actions related to a product line system are vital to a line's success. Complexity theory suggests that this is true for all complex systems. We also know that the tools, practices, and models, the system enablers, that work well for one archetype might not be as effective for others. It's important to recognize that when a company "transforms" a product line (changes the archetype), the current patterns of behavior and practices used by contributors may hinder the line's performance.

An anti-pattern is a behavior exhibited by system agents that hinders good performance and leads to poor outcomes. It acts as kryptonite to a product line's ability to evolve. This observation doesn't mean you should eliminate the components or agents causing the anti-pattern. Instead, it suggests that teams should address negative behaviors by offering incentives, influencing change, and implementing direct adjustments that promote desired behaviors. This requires companies to modify specific system agents, practices, tools, templates, and models. This insight forms the basis of the Product Line Change Theory, which I discuss in Chapter 22.

Mapping Change

A product line roadmap should depict a team's ideas about improvements and modifications to the key parts of their line, including platform-levers and the line's archetype. A system assemblage map illustrates the team's approach for related enhancements and adjustments to the system's non-key part

162

agents. An archetype change, considered a product line transformation, will be marked on both maps.

Understanding the connection between an archetype and the system's behavior patterns is crucial for effective product line systems management. And taking that a step further, we see that mastering product line archetype congruence with a business model is a skill most teams and organizations should develop if they want to stay in the Accelerating Change Race.

Archetypes and The Accelerating Change Race

Product line teams must understand their system's archetype and the behavior it induces. This knowledge is essential for enhancing your line's performance and achieving significantly different outcomes.

When chain-link functions and external systems display behaviors that contradict the system, it is important to understand these behaviors in relation to your product line system's archetype and the business model in which it operates. Several things will need to change. Overcoming this misalignment begins by viewing it in relation to a system's archetype.

Chapter 11 Summary

System Archetypes, Business Models, & Strategy

- **An Architecture's Basic Form:** A product line system's archetype is the line's architecture's basic form and starting point. The archetype defines how the system works.

- **Platform-lever Types Determine the System's Archetype:** A product line's platform-levers are the foundation of the system's "archetype." Changing platform-lever types changes the system's archetype.

- **Platform-levers, Archetypes, Strategy, and Business Models:** A product line system's archetype, which is determined by a lines platform-levers, is foundational to its strategy and the alignment with the business model in which it operates.

- **Archetypes in Action:** Tesla upset the auto industry using new platform-levers: software, batteries, and manufacturing. When launched, these platform-levers created a system archetype new to the automotive industry, which caused notable disruption.

- **Tesla's Systemic Advantage:** Tesla's advantage lies in effectively managing its new archetype. Other automakers face major difficulties in catching up. The competitors, burdened with debt and behind in innovation, are too late to respond effectively to Tesla's success.

- **Competitive Landscape in EV Market (2024):** Traditional automakers like Toyota, Ford, GM, and VW face challenges catching up with Tesla's unique system archetype. They struggle with high costs, delays, and a lack of understanding of the full product line transformation. The industry's mature competitors struggle to improve the Outcome Velocity of their EV product lines.

- **Actor Behaviors and System Enablers:** When the system archetype changes, the ways people work and make decisions may no longer match the system. New behaviors and enablers are needed.

- **Critical Knowledge:** Teams must understand their product line system's archetype, plus its actors and enablers. How well all agents work in unison will determine the line's Outcome Velocity.

164

Chapter 12

Archetype Disruptor Theory

System Change That Disrupts Industries

When we change platform-lever types, we also change the archetype. However, a compelling strategic move occurs when a company develops a platform-lever type that is new to an industry. Such uniqueness is essential for disruptive and radical innovations.

In 1995, the late Professor Clayton Christensen introduced his theory of Disruptive Innovation.[52] These types of innovations can disrupt and transform industries. Uber is an example of a disruptive innovation in the Taxi industry. A quick online search reveals many more industry disruptions.

Christensen never addressed disruptive innovation from a systems perspective. However, his theory explicitly states that disruptive innovations come from small start-up companies. He reflected the outdated belief that large, established companies don't change and that technological shifts mainly happen through small firms.

Disruptions by Large Companies

Radical Innovation was introduced a decade before Clayton Christensen's Disruptor Theory. However, a few years after Christensen's work, Gina O'Connor and her colleagues at Rensselaer Polytechnic Institute (RPI) expanded the concept of "Radical Innovations" at large companies.[10] The difference between Disruptor Theory and Radical Innovation lies in the size of the company. As the RPI team explained to us, this difference is clear in behaviors and forces found in big companies but not in small ones.

Over the past decade, innovation experts have shifted the term "Radical Innovation" to "Breakthrough Innovation," giving it a more positive rather than rebellious tone. However, it still shares the same core novelty as disruptive innovation, but it applies to large companies instead of small ones. When someone talks about disruptive innovation by large companies, the discussion usually revolves around breakthrough innovation. Despite this, many of us still use the term disruptive even when referring to large companies.

Until now, those involved with disruptive and breakthrough innovations have not focused on product line systems. No one recognized or understood the archetypes of product line systems. Instead, the single-product paradigm dominated everyone's thinking.

The Archetype Disruptor Theory

Building on my understanding of product line archetypes, experience, and system awareness, I have an important theory to share. I call it "The Archetype Disruptor Theory." It's that:

Industry disruptions occur only when innovation originates from a system archetype that is new to the industry, regardless of the company's size. Also, for new system archetypes to succeed and expand, they must align with the corresponding business models and strategies.

This theory proposes that a large company's innovation can only disrupt an industry when it introduces a new product line system archetype. Note the importance of viewing disruption from a complex, multi-product system perspective rather than a single-product, orderly system view.

New to an Industry, Not the World

Disruption doesn't happen simply because a system archetype is new to a company; it only occurs when a system archetype is new to an industry. Contrary to what Christensen and his followers argue, the size of the originating company doesn't matter.

Some innovation experts might challenge this statement. Its proof comes from analyzing many products and innovations labeled as disruptive or breakthroughs. I include a summary of this work in the online appendix. The results indicate a strong link between a platform-lever's uniqueness in an industry and disruptive innovation. Plus, I have not found successful breakthroughs or disruptive offerings with an archetype that is not unique to its industry.

Improving efficiency in executing work, making decisions, and fostering creative thinking can boost a product line's performance when competitors share a common archetype, but it won't lead to disruption. The reasoning is straightforward. When competitors offer similar product line archetypes, they can easily copy each other's strategies. However, when a distinctly different archetype enters the industry, creating effective competitive responses becomes difficult. Refer to Case Study 2 in the appendix on Johnson & Johnson for an example. You'll see how Johnson & Johnson's new archetype for their contact lens business eliminated multiple competitors.

Other Factors Also Matter

Not all new archetypes in an industry will cause disruption. A line's performance and speed of results depend on many interconnected system parts and agents. Market needs, segmentation, technology building blocks, and system behaviors are also crucial.

Remember, no single factor, not even the archetype, is the sole determinant of the system's behavior or outcomes. Consider Airbnb. This company's novel archetype did not disrupt the hotel industry. It also did not solve most of the industry's Jobs-to-be-Done clusters (Verb-based segments). However, it did create a new segment within the industry.

Several other system obstacles must be overcome for a unique system archetype to truly disrupt an industry. If the innovation fails to clear any of these obstacles, it might be considered novel, but it won't be a true disruption.

1. The system archetype must align with the business model in which it operates. (Product Line Congruency Theory)

2. The offering(s) must meet customers' Jobs-to-be-Done at a sufficient volume to justify its commercialization.

3. Price points and costs can be set to ensure the system generates a positive cash flow.

4. The technology must be ready and stable enough to ensure acceptable product quality.

5. The operational plan must be designed to overcome consequential bottlenecks and constraints. This is a challenge for orderly systems practices and methods.

6. The system and its strategy present a competitive moat that prevents other organizations from replicating and undercutting the product offerings.

7. The targeted strategic moves have a reasonable chance of enhancing the line's Outcome Velocity (increasing cash flow, customer satisfaction, and competitiveness).

8. There are no impossibilities or "game-stoppers" in the product line roadmap or system assemblage map.

A Game of Archetype Disruptor Chess

Electric vehicles remain a dynamic arena for archetype disruption. New technologies, such as solid-state batteries and processes enabled by artificial general intelligence (human-level) automation, continue to reshape the auto industry's what, how, and why.

Be aware that significant changes are necessary when the archetype of a product line system shifts. Both the line and its parent company must adapt the product line system's archetype. The organizational structure, processes, practices, and strategy all need modification. These adjustments explain Tesla's employment fluctuations throughout 2023 and 2024. The rapid and continual archetype change that Tesla adopts is not suitable for employees seeking a stable lifestyle.

Tesla's approach renders the idea of industry disruption outdated. They serve as the first example I've seen of constantly evolving archetypes. Tesla normalizes changes in product line archetypes, making them the norm rather than an exception. Such rapid changes will probably become standard in the Accelerating Change Race.

Disruption Everywhere

Large language models, combined with industry-specific knowledge, generative AI, reasoning, and self-learning AI, will enable teams to create innovative, industry-first platform-levers. This will introduce new system archetypes across various sectors, leading to significant disruption. Many experts anticipate that AI's rate of improvement will be at least four or five times annually, with compounding effects each year. Over a decade, this results in more than a millionfold increase in performance. Such performance gains and their impact on product line systems are unpredictable. Still, it's certain that future product line systems and their archetypes will look very different.

Every product line team should aim to understand system archetypes and their impact on disruption. Their success will depend on it.

Archetype Disruptor Theory and Accelerating Change

Understanding Archetype Disruptor Theory is essential for keeping up with rapid change and will remain so in the future. In every industry, shifts in archetypes will become the norm. The issue is not whether it will happen, but when and how it will appear. No product line team should want to find itself facing a new-to-their-industry system archetype. And every team should aim to position their competitors in that situation.

The Archetype Disruptor Theory offers a crucial framework for understanding how industries undergo transformation through system-level innovation. The theory challenges traditional views on disruptive innovation by emphasizing that genuine industry disruption is not determined by a company's size but by the introduction of a product line system archetype that is completely new to that industry. Unlike perspectives that focus on single-product and orderly systems, the theory proposes that industry disruption comes from the uniqueness of the product line systems architecture, representing a new archetype within the industry.

The key insight is that an innovation's disruptive power stems from its uniqueness within the industry context; when a new system archetype emerges, it presents competitive challenges that incumbents struggle to meet due to their existing product line system architectures.

Illustrative examples, such as Johnson & Johnson's transformative approach in the contact lens market and Tesla's ongoing dynamic shifts in the electric vehicle sector, demonstrate that success goes beyond just having a product line archetype that is new to an industry. It also requires that companies overcome key challenges related to market needs, operational readiness, and maintaining a competitive moat.

Ultimately, the Archetype Disruptor Theory redefines industry disruption by asserting that innovation must reshape the multi-product system to create notable change, emphasizing systemic transformation over mere performance improvements.

Chapter 12 Summary

Archetype Disruptor Theory

- **The Archetype Disruptor Theory:** The theory introduces the concept that industry disruptions occur when a company leverages a product line system archetype that is new to the industry—not merely through radical innovation in isolated products, but via a system architecture change.

- **Beyond Single-Product Innovation:** The theory challenges the conventional single-product paradigm by revealing how the architecture and specific archetype of a complex multi-product line system, when introduced to an industry, can disrupt that industry.

- **Size Is Not the Deciding Factor:** Contrary to traditional views (such as Clay Christensen's Disruption Theory), industry disruption is not dependent on the size of the company. Instead, what matters is introducing an archetype that competitors have not encountered in their industry.

- **Critical Success Factors for Disruption:** Successful disruption requires more than novelty; the new archetype must align with market needs, utilize robust technology, be operationally feasible, and have a viable business model. Together, these factors create a competitive moat.

- **Real-World Illustrations:** Examples such as Johnson & Johnson in the contact lens market and Tesla in the automotive industry underscore how unique new archetypes can disrupt established competitors and reshape entire industries.

- **Proactive Response to Emerging Archetypes:** Product line teams must work vigilantly to monitor and react to emerging system archetypes, ensuring they are never caught off guard by new system archetypes and are prepared to respond when they arise.

Chapter 13

Navigating & Exploiting Constraints & Forces

Defining Context and Maximizing Outcome Velocity

Before exploring the world of constraints, it is essential to recognize that all complex systems have constraints. If a system survives beyond the short term, it does so by navigating past, over, or through these constraints. Another perspective is that constraints shape the architecture and coherence of successful systems. Only those architectures that withstand system constraints can be deemed successful. The question is whether future constraints might pose insurmountable challenges based on the system's architecture.

Internal and external obstacles and events create an array of constraints that affect strategic moves, the system's flows, and a line's Outcome Velocity. Root cause analysis experts appreciate hearing about such constraints. Unfortunately for them, complex systems issues do not result from a single root cause. Instead, teams need a different approach to complex product line systems to address such constraints and improve the system's flows.

Constraints Define the System Context (Substrate)

Constraints also relate directly to the context in which the system operates. If the context of the system changes, so will the constraints. Professor Alicia Juarrero explains this in her 2023 book "Context Changes Everything: How Constraints Create Coherence."[53]

In Chapter 2, I highlighted the differences between multi-product management and the typical single-product context in which most companies operate. This means the constraints teams must consider for multi-product systems differ from those in the single-product context.

The coherence that Professor Juarrero refers to is the behavior of the complex system. A product line's coherent behavior arises from aligning the system architecture with its constraints. Understanding the constraints that limit a line's performance in a multi-product context provides a team with a foundation for assessing their line's architecture. The system's design—its key components, actors, enablers, and other agents—must work coherently against the constraints. As you conduct this analysis, several questions quickly arise.

- Are you sure you have identified the important constraints?
- Can you alter or change any constraints?
- What type of constraints must the product line team address?
- Which constraints might cause hard stops to your system's flows?

- Which constraints are permeable, and which are flexible?

- Are the constraints interdependent? Do they cluster or relate to other constraints?

- Which aspect of your line's Outcome Velocity— customer satisfaction, competitiveness, or cash flow—is hindered by each constraint?

- How much effort, money, and time would it take to alter each constraint?

Every product line team's most important job is to understand the forces and constraints impacting their line.

Forces and Constraints

Where there are forces, there are constraints. And where there are constraints, there are forces.

Product line management requires teams to understand, respond to, and create influences that affect the line's Outcome Velocity. For example, developing products and services that customers demand is a positive force builder. However, constraints always exist that work against or limit this force. If they didn't, it would be like discovering free room-temperature superconductivity.

Constraints may be operational, such as production bottlenecks; financial, like budget concerns; or related to supply chain issues, including concerns about raw material quality. Any factor affecting the product line in each product's value chain may involve a constraint. The potential for constraints appears limitless.

To spot the constraints on any product line system, examine its flows—work, information, decisions, and outcomes—and its recent past and future Outcome Velocity dimensions, customer satisfaction, cash flow, and competitiveness. Any factor that hinders or limits these streams is a constraint.

The Constraint Map

While some forces and constraints are common to all players in an industry, each product line also has unique forces and constraints that teams must deal with.

Thankfully, complexity theorists have developed an approach for tackling this task. It's to set up a map, a playing field, of the line's constraints.[54]

Teams can conduct constraint mapping as a one-time event, but its value increases when established as an ongoing process. It should run concurrently with Product Line Roadmapping (focused on advancing the system's key components) and System Assemblage Mapping (focused on enhancing system agents and architecture).

Figuring out a product line's constraint map involves not only breaking down the constraints but also creatively and strategically recombining that breakdown into insights for mitigating or overcoming the constraints.

This work requires deeper thinking and analysis than it may seem at first glance. We recognize that constraints are not equal. Some are substantial; others are minor. Some necessitate a considerable amount of time, resources, and staffing to address, while others can be relatively straightforward to handle. Certain constraints are easy to identify, while others might be concealed within an organization's culture and consumer behaviors.

I encourage all teams to embed Constraint Mapping into their system as an enabling agent. Learning about Estuarine Mapping through the teachings of complex systems expert Dave Snowden is a great starting point.[56] It provides the most thorough examination of complex system constraints and a means to plot them based on the energy (effort, money, and resources) and the time needed to mitigate each.

Not So Simple

Consider how difficult it can be to identify and expose certain constraints for everyone to see. Success hubris, for example, is a significant issue affecting product line changes. Just because an organization succeeded in the past (as all large companies have) doesn't guarantee success in the future. Nor does it imply "we know best." Yet that is precisely the mindset that some leaders apply when attempting to improve a line's Outcome Velocity.

Success hubris is a substantial constraint that redirects the flow of outcomes. In Online Appendix Case Study 1, I present a case study of Keurig Coffee, which urgently required a new product to compensate for its single-serve pod nearing patent expiration. The case study illustrates how the company's leadership's hubris hindered their product line from making coherent moves. Identifying and addressing success hubris can be challenging, particularly when leaders have achieved remarkable success.

Other constraints can also be challenging to identify. For instance, a natural disconnect arises among team members when a constraint seems outside their capabilities. The constraint may appear so far beyond their reach in terms of size, scale, and cost that there's little incentive to invest time in figuring out how it might be mitigated. That's a problem when competing against aggressive companies like Tesla. These factors don't deter this company; they welcome them. They understand that traditional competitors are reluctant to embark on significant constraint-busting initiatives. The first to do it stands a chance of disrupting an industry.

Tesla's mega battery plant, which targets fundamental cost and supply constraints, has far surpassed the investment that any other US automotive company would consider. The company's mantra seems to be "Do it big, do it right." In contrast, other firms appear to embrace "Do it small, avoid risk." This latter approach does not bode well in the Accelerating Change Race.

Some constraints are easy to identify and respond to. These often relate to engineering and regulatory issues. Often, the challenge lies in being completely unaware of the constraint. It's not uncommon to see groups of intelligent individuals try to address a constraint while overlooking key factors. The laws of science, customer Jobs-to-be-Done, social impact, and skill requirements are among the most significant. Unfortunately, those who lack awareness of these facts are oblivious to the constraint. Regrettably, this lack of knowledge can affect even the best of us.

Plot The Forces

Evaluating the impact of each force on the system can be challenging. Refer to Table 13.1 below. There is always significant uncertainty because product lines are complex systems. When we address one force, another will emerge in its place. A team's goal should be to make progress against the force and promote positive emergence.

Table 13.1 (same as Table 9.3 in Chapter 9): Key Forces Affecting a Product Line's Assemblage.

Forces	Discussion
Needs, Wants, Desires | Jobs to be Done Outcomes	These forces attract customers to each product. Customer needs always exist, but the force doesn't come into play until a set of product attributes is available.
Disruptions (slow); Black Swan Events (fast); Weak signal insights.	Other independent forces can disrupt all internal and expected forces.
Regulatory Demands	Regulatory boundaries can shape customer demand and attribute availability. Teams usually know the regulatory demands on their products, sometimes based on disruptions.
Risk/Threats (Real & Perceived)	We may view risks as the likelihood of a negative event affecting the system. One problem with risks is when perceptions drive behaviors, causing negative forces to appear.
Superiority, Quality, and Value (Real & Perceived)	The customer's perception of products is a notable force. How well the products match the customer's beliefs about the superiority, quality, and value of each offering reflects a force that either helps or hinders a line's Outcome Velocity.
Attribute Positioning | Matches, plus Synergies	Attribute positioning (engineering) and perceptual positioning (marketing) affect customers' attraction to the products. Goal: Engineer a product's attributes specific to Verb-based segments (clusters of JTBD outcomes). Target through engineering before communicating positioning through marketing. See Appendix 7 for more on attribute positioning.

Forces	Discussion
Competitive Push, Moves, Blocking	Competitors create forces that work against your offerings in the battle for market share.
Leverage and Scale, Flexibility	Consider Wright's law of scale and cost reduction; increasing delivered value; easing future system changes; augmenting the offerings; and maximizing velocity in the near and long-term.
Organizational Synergy/Alignment	The combined effect of parts and forces is greater than the sum of their separate effects.
Organizational Inertia	An organizational entity will continue its current course until a sufficient force causes it to alter its direction.
Resource Capabilities & Capacity: Assets, Money, Skills, and Talent	Having the right resources and the capacity to deploy them into projects can be a positive force toward improved product line Outcome Velocity. Not having them can be a negative force.
Complexity Costs and Overhead	As more products and product variants are added to a product line, the total overhead cost will increase disproportionately.
Brands (both positive and negative)	Brands and brand loyalty can be a powerful force that affects a product line's offerings and connection with customers. Strong competitor brands can be negative forces working against a line.
Diffusion Dynamics	The adoption and acceptance of technology or products by customers and markets. Understanding the forces that induce or hinder diffusion aids in proximate forecasts.

There must be evaluation methods for the system forces, their interplay, and scenarios regarding the possible consequences of decisions and actions. Teams can improve their judgments by seeking input on the most important factors and considerations. Ultimately, the team leader must take ownership of all decisions. The hope is that all decisions, judgments, and choices will be based on insights and principles, not consensus. Argue as if you're right but listen as if you're wrong.[55] Throughout the book, I discuss insight generation and PLSM principles. Chapters 17 and 23 address these topics directly.

Force-field maps complement the parts-oriented roadmap and the constraint map. These mapping activities provide contributors with a clearer understanding of the forces that influence outcomes. Most importantly, they shift the focus from figuring out how to expedite development projects to prioritizing the improvement of the product line's Outcome Velocity and addressing both efficiency and change challenges.

The mapping work helps teams develop a shared understanding of the line's dynamics. Figure 13.1 presents an example of a simple force-field map related to a strategic move.

The magnitude and direction of the forces, along with the confidence and agreement with these judgments, are key considerations for product line planning. Insights gained through this work can contribute to deep discussions about current and potential parts and agents.

Late Moves

One key to improving Outcome Velocity is to start strategic moves before their outcomes are necessary. It may sound insultingly simplistic, but it is an enormous challenge for a product line team. Delayed actions can undermine product lines.

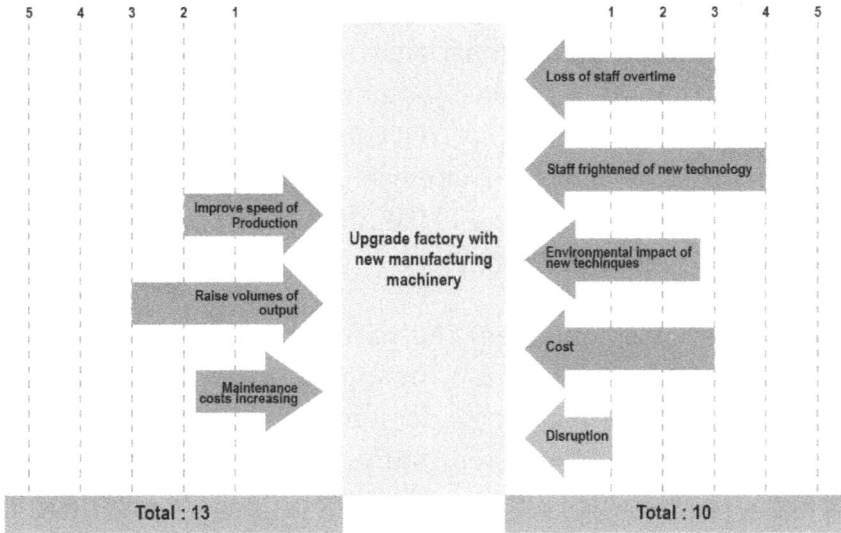

Figure 13.1 *Most companies create ad hoc or temporary teams to address significant opportunities and challenges throughout a product line's lifecycle. This often occurs when FOBO (fear of being left out) affects a leadership team. When you notice this happening, consider it a sign that you may already be late.*

FOBO responses for an existing product line directly relate to the tug-of-war between efficiency and change. Some managers aim to operate more efficiently, while others pursue change. This brings up the question: Should a strategy focus on leveraging existing strengths, addressing weaknesses, or both? When we concentrate our efforts on efficiency and change, you'll often find that the answer is "both."

Early Outcome Velocity improvement and constraint exploration can be powerful when conducted by intelligent individuals with the right resources. The goal is to identify strategic moves to enhance the line and prevent delays. First, you need to gain insights into Outcome Velocity, constraints, and the components and agents of the system.

Beware of Strategic Debt

Heavy demands for near-term outcomes can delay long-term initiatives. When teams decide to defer long-term strategy moves, it's important to do so deliberately and transparently, while maintaining a record of the deferred actions. Teams accumulate these postponed or overlooked strategy moves in a "Strategic Debt" bucket, like an Agile backlog.

The Strategic Debt bucket plays a vital role in breaking the Negative Innovation Loop. Eventually, you will need to pay off this Debt. You won't understand the severity of your challenge unless you accumulate and report it.

Strategic Debt is a hidden negative force that can turn against daily operations at any moment. It represents a threat that can instantly derail your roadmap. You will always incur Strategic Debt. The higher the Debt, the riskier the situation.

The Resiliency Formula

All product line teams need a sense of "urgency before an emergency."

Every contributor must take action when unexpected problems impact a line's Outcome Velocity. Some managers may choose to wait for leadership to respond or hope that the system's adaptive nature will engage and miraculously resolve the issues without their intervention.

The speed of responding to an unexpected event can be just as important as the actions taken afterward. However, situations may require individual contributors, not just top leaders, to make quick decisions and act immediately. For example, a fire or a malware attack could threaten an entire product line. In these cases, waiting for an executive team to decide what to do could cause serious damage.

Rapid Resiliency for Product Lines

Responding to an unexpected black swan event is challenging, but necessary. When the speed of the response is critical, it's important to delegate decisions and actions to the lowest levels of staff and management. Teams should establish a fallback or rapid approach to facilitate quick responses. You can't afford to waste time by escalating issues to higher management. All contributors must understand these steps before the unexpected occurs.

1st: As best you can, avoid getting killed or harmed by unexpected events.

2nd: Observe, be aware, and build insights about what's happening.

3rd: Quickly re-prioritize strategic moves purposely to mitigate major issues. A response to the unexpected event may supersede all strategic moves underway.

4th: Communicate and collaborate with all actors as well as internal and external ecosystems. Emphasize the absolute need to communicate actions, choices, and insights. Collaborate and coordinate all actions. Seek to maximize decision coherence and work coordination.

5th: Create and carry out new moves based on new insights. Recast the portfolio of strategic moves to prioritize near-term survival over long-term well-being.

Making the Moves

Product line management involves creating strategies in response to external forces and internal constraints. However, these strategies are ineffective unless a team can effectively carry them out. This might sound obvious, but every product line team will inevitably seek shortcuts around actions that were once deemed important.

Navigating Constraints and Forces and The Accelerating Change Race

Competing in the Accelerating Change Race requires teams to maximize the Outcome Velocity of their product lines in the near, mid, and long term. The job will demand both efficiency gains and adaptability depending on the situation. To achieve this, teams must adjust and manipulate the components and agents of their product line system in relation to constraints and forces.

Three issues stand in the way of success. Consider these questions:

- *Does the team understand all system components, constraints, and forces?*

- *Can the team form actions to navigate and exploit the constraints and forces to improve the line's Outcome Velocity?*

- *Do the actions in one time horizon constrain outcome emergence in another? Does that matter?*

Constraint mapping and navigation serve as powerful tools for enhancing efficiency and driving change.

Once a team identifies and evaluates their system's constraints or forces, they should establish a path forward to address or circumvent them. If certain constraints frequently arise, the team should develop or refine enablers to monitor and mitigate them.

Chapter 13 Summary

Navigating Constraints

- **Forces on the Product Line:** Internal and external changes are significant forces on a product line. These forces require teams to make strategic moves to accelerate positive outcomes and mitigate negative ones.

- **Constraints:** All systems always have constraints, some are hard stops, some are pliable, and others are porous. A team's job is to remove, mitigate or navigate around critical constraints. When one is dealt with, another will surface.

- **Hubris and Strategic Debt:** Success hubris, the belief that past success guarantees future success, can hinder and stall needed product line changes. Teams must begin strategic moves before they become late. Late strategic moves increase the line's Strategic Debt, making future moves more difficult.

Considerations from Experience:

1. Teams must envision new technologies as potential platform-levers, rather than as building blocks.

2. Discrete technologies contributing only to one platform-lever should be part of that platform-lever, not as a separate technology building block.

3. Verb-based segments are Jobs-to-be-Done clusters. They are targets at which teams should engineer attribute sets (products). Teams should embed details of Verb-based segments as line themes in Product Innovation Charters (PICs).

4. Response to unexpected events requires rapid resiliency. Resilient responses include observing, re-prioritizing strategic moves, communicating, and casting new strategic moves.

Chapter 14

External Systems & Performance

Chain-links and Ecosystems

In Chapter 6, I discussed how workflows, cash flows, and decision flows intertwine to enable the outcome flow of a product line to emerge. A product line's performance depends on each flow progressing close to unimpeded. However, we also know that other systems, separate from the product line, impact the line's outcome flow.

The COVID-19 crisis revealed the importance of external systems. Supply chains affected nearly every business, and many product lines had to respond to difficult situations. Yet, with support from governments and communities, these external systems adjusted quickly and en masse. The fear of what would happen without adaptation was palpable.

Many systems, including the product line, make up a larger ecosystem. These systems, both inside and outside the organization, influence the performance of the product line. Each non-product line system has its own structure and processes. Each is also participating in the Accelerating Change Race and must be responsive and resilient as needed.

For product line teams to achieve the best possible outcomes for their line, they need to develop capabilities that can influence external systems and their related flows. They must understand and engage with all external systems, and those managing each system will have their own views of what's best for their system.

Flow Resonance and Outcome Emergence

In large companies with numerous products, many individuals are responsible for various processes impacting the line. This also applies to external systems with workflows intended to align with the line's processes. The challenge lies in getting everyone to collaborate effectively to enhance the line's output velocity. This is a demanding task.

Consider how workflows and decision flows across external systems may orient in different directions: one emphasizing efficiency and another seeking change. You might observe a group or an actor within an external system using software that fails to streamline data and analytics in conjunction with a product line system's workflow. Alternatively, the information flow in an external system may not correspond with the product line's system archetype. In such instances, the information flow isn't streamlined and does not align coherently with the product line's workflows and decision flows. When these flows don't match, the line's outcome emergence will slow down or stall. This mismatch adversely affects the product line's Outcome Velocity.

Flow mismatches across external systems can make achieving resonance—the boost beyond alignment—nearly impossible. Resonance occurs only when the flows converge toward common efficiency gains and transformative outcomes. If one flow diverts efforts in the wrong direction, discord rather than resonance is more likely. (see Chapter 7)

Maintaining flow resonance within the product line system and across external systems is one of the biggest challenges product line teams must face. It's central to product line systems management and becomes more difficult as change accelerates. Consider the problems if a supplier informs you that they will no longer produce a key component of your platform-lever. Or think about the issues you might encounter if your IT department notifies you that they no longer support your product development software, especially if it's not approved by SAP or Oracle.

The Internal Chain-Link System

The most important systems are external to the line but internal to the business. These systems are established by each function within the business unit. For example, finance, HR, IT, operations, supply chain, sales, and marketing have systems in place to coordinate their respective flows. Consider each system a "chain-link" in the business unit's "chained" strategy.[56] See Figure 14.1.

Each function has its own strategy, decision flow, workflow, information flow, and outcome flow. As strategy expert and UCLA Professor Richard Rumelt points out, poor performance or misalignment of any link can weaken the entire chain. However, more importantly, the alignment with and potential boost from non-product line links affect a product line's Outcome Velocity.

Figure 14.1 A view of the Chain-Link Business Strategy analogy. Each functional strategy is a link in a chain that creates the full business strategy. A Product Line Strategy is just one business unit strategy chain-link.

191

Achieving alignment and resonance across chain-link flows presents a greater challenge than ensuring flow coherence within the product line system. Nevertheless, from the business unit's perspective, the product line system is simply another link in its chain-link strategy. To facilitate a match, the product line and chain-link (functional) flows may require adjustments. Such flow alignment across the chain-link functional groups is the purpose of business models.

When the internal links align, the business unit strategy should function smoothly. However, this strategy will stumble if the alignment of the chain-link and the product line becomes disjointed.

Chain-links, Archetypes, and Business Models

It's essential to understand how the business model shapes the interplay among chain-links. The business model dictates how functions should interact, not in detailed specificity, but rather as guidance toward desirable outcomes. If you change the business model, you must also change the interplay among the functional chain-links.

Yet, if we change a product line's archetype, we also alter the system's behavior and workflows. This relationship indicates that the external chain-link flows must adjust to support the new product line archetype. Today's business management highlights the importance of aligning functions with a business model. However, we do not find inquiry or response to a product line system's archetype in business model planning or on the business model canvas. This is not surprising, since business models and the Business Model Canvas predate product line systems management by more than a decade.

From a systems perspective, the relationship between the product line archetype and the business model, along with the chain-link strategies, is a cornerstone of the line's performance. If we change the line's archetype, we must adjust the line's strategy, business model, and chain-link strategies. Let me offer this hypothesis:

The product line archetype determines the business model and the chain-link strategies. If you change one, you must change the other two.

This principle also means that a business model and chain-link strategies cannot be optimal for two different product line archetypes. We see this happening all around us.

Consider how mature auto companies struggle to develop electric vehicles (EVs) alongside internal combustion engine cars. Or a commodity chemical company finds it difficult to build a specialty chemical line or a new service. When companies develop product line stacks that incorporate technologies forming new platform-lever types, they are also changing the system archetypes. This change then requires new business models and revised chain-link strategies.

Boosting Outcome Velocity: Chain-Link Innovation

We know that services and financing boost the value of automobile product lines, just as apps and content add value to cell phone product lines. These boosts extend beyond simple alignment with products. To achieve these gains, teams must foster creative and strategic thinking within the non-product line chain-link functions and collaborate on implementing potential innovations. Such cross-chain-link innovation can significantly increase a product line's Outcome Velocity.

I share three case studies on internal systems—chain-link strategies—in this book and the online appendix. Each case demonstrates how non-product line functions enhanced a product line in ways the product line team could not achieve otherwise.

- **Narrative 1: Online Appendix Case 1.** GE's Plastics Sales Methods and New Polymers (Jack Welch's big success). GE Plastics bundles a new polymer offering with a service (two separate platform-levers) to open potential markets. Dow went further. They incorporated molecular modeling as a platform-lever in their new olefin product line.

- **Narrative 2: Book Appendix Case Study 4.** Apple's Potent Chain-link Strategy. It's not just the products. The chain-link includes Apps, the Retail Store, Intellectual Property, the iBrand, Apple Silicon, plus other functional links. Together, they create a formidable chain. Can any competitor possibly beat them?

- **Narrative 3: Online Appendix Case 2.** Lennar Homes introduced a radically new platform-lever into an unchanged business model.

The major challenge in achieving flow resonance across chain-links arises from traditional organizational structures and how these management approaches can separate and isolate roles, not just functions. It's natural to perform a role by staying within set organizational boundaries. Those in decision-making roles remain in the decision flow stream, while those who design and create keep to their workflow stream. Meanwhile, those who set up IT tools stay within the information flow.

Influencing Chain-links and Business Models

Typically, you'll find that no individual has the responsibility or authority to alter all of the workflows across the chain-links. Nor is anyone tasked with driving innovation across links. A product line leader is more likely to say, "I can change our workflows. I can influence a change in our information flow. However, the decision flow also involves senior managers. And, as for chain-link innovation, I'll discuss it with others. But don't expect a good reception to carry out cross-organizational work if it demands that these functions change how they do things."

Such chain-link changes must be the responsibility of a governance team. They are responsible for overseeing its relationship with the business model and the various chain-link flows. Although product line teams must report necessary changes to a governance team, the product line team cannot independently alter the business model or change the chain-link strategies and flows.

These top managers must pay special attention to the flow and velocity of outcomes for the line. You'll find more on governance teams in Chapter 23.

External Systems

Some systems that affect a product line are external to the company. These external systems also have structures, actors, and enablers. Similarly to internal chain-links, the product line team must strive to align with and derive meaningful benefits from the external system.

External company systems can be categorized into three buckets.

- Those providing or facilitating input to the product line

- Those receiving or facilitating output from the product line

- And those that guide or regulate any system, including the product line

Input Provider Ecosystems

The actors in this external system include suppliers and partners. Suppliers provide the raw materials or discrete system parts for the line, while partners may offer services and support that benefit the line. For example, software products often utilize APIs and webhooks to integrate with third-party software without customers being aware of this integration. To customers, it's one seamless product. However, opportunities and challenges arise when the product line's performance relies on an external ecosystem.

The company that manages the product line often seeks to handle these systems through departments or functions dedicated to supply chain and partner management.

Outside In

Over the last decade, there has been a trend for product lines to integrate supplier and partnership roles into a company's internal system. For decades, many companies have pursued efficiency by outsourcing and offshoring supplier functions. Undoubtedly, this separation from suppliers and partners undermines a product line's ability to change and affects the team's resilience. Now, however, the focus is shifting from outsourcing for efficiency to internalizing for change resilience.

Consider how Apple developed its own internal CPU chip-making capabilities to replace Intel as a supplier. Apple Silicon, an internal business unit, aligned development with Apple's operating systems (M chips + iOS) to provide significant leverage within its computer and tablet product lines.

Ford CEO Jim Farley recently discussed his company's challenges in adopting new technologies and collaborating with suppliers during an interview.[57] He compared Ford's struggles to how easily Tesla can make changes. Farley explained that Ford's emphasis on outsourcing to cut costs and enhance efficiency had been a strength for the company. Nevertheless, it has left them unable to adapt quickly to the demands of electric vehicles. An auto industry expert cautions that Ford's intense focus on efficiency could now represent a weakness and potentially lead to negative consequences.[58]

Other established auto companies face challenges similar to Ford's. Each automotive product line requires significantly greater leverage and enhanced capacity for change—they must both perform and transform. Their typical strategy of outsourcing for efficiency contradicts the need for Change Capacity in the line.

There is much to learn about external ecosystems. As you explore further into how they operate and how to benefit from them, consider reviewing Paul Hobcraft's insightful work.[59] Hobcraft has worked extensively on ecosystems and innovation.

Outcome Emergence and Verb-based Segments

Both internal and external ecosystems also encompass means to address markets, customers, and distribution channels. The product line team has no authority over these entities.

A major issue for product lines is that internal and external systems often organize customer-facing sales and market management to align with Noun-based market segments, such as geographies, industries, and demographics. Yet the most crucial alignment is between a product line's offerings and Verb-based segments. This suggests a potential mismatch between a Noun-based focus and Verb-based customer needs. In such cases, challenges may arise that might hinder a line's Outcome Velocity. I provide more details on the difference between Noun and Verb-based segments in Appendix 7.

Other mismatch issues may also arise concerning branding, pricing, and service. These issues vary for each product line but become significant as they aim to position products uniquely within different market segments. Establish a strong match, and you'll enhance your line's Outcome Velocity. Handle it poorly, and you'll stall the Outcome Velocity.

Output Recipients

Nearly all industries have regulatory bodies and standardization groups, and product lines must comply with the rules and laws established by these entities. Significant mismatches occur when a product line includes parts that fall outside the established boundaries. In some government-regulated industries, such boundary crossing can be detrimental to the line. This situation can be observed in various sectors, including medical devices, pharmaceuticals, FinTech, chemicals, and certain high-tech industries.

Consider the challenges associated with FDA regulation for medical devices in the United States. The mechanisms and infrastructure for reviewing these devices stem from an outdated single-product paradigm rather than a multi-product system. Each product in a line necessitates separate clinical testing. There is no straightforward way to obtain FDA approval for a platform-lever or a new archetype.

I don't want to imply that I know how to conduct platform-lever clinical studies, especially since platform-levers must continuously evolve. I'm simply highlighting that we lack a safe and efficient path forward for an entire product line system. Consequently, regulated ecosystems do not fully capitalize on the economic benefits of product lines in the medical device industry. The same applies to other regulated industries and product types.

Industry standardization bodies are crucial entities in a line's external system. However, they are distinct from government regulatory bodies. It's not illegal to deviate from industry standards. Nonetheless, doing so may cause your line to conflict with an ecosystem. The main advantage for existing product line teams is their participation in standardization committees that could impact their line's future. There are hundreds of standardization entities across the world.[60]

When product lines introduce new-to-the-world platform-levers (a new system archetype), such as we see today with artificial intelligence, there is rarely any regulating body or standardization group to set boundaries. This situation will change as technologies mature, but capitalism's free hand forms the only limit. If a team manages a new-to-the-world platform-lever, they must work to engage with and potentially assist in developing the standardization bodies.

Deliberate Resonance and Boost

Product line teams should strive for flow resonance with all external and internal ecosystems that influence the line's performance. Unfortunately, waiting for this to happen spontaneously resembles a 'Waiting-for-Godot' experience. The problem is that poor alignment and a lack of cross-link innovation increase the likelihood of subpar performance and slower outcome velocities.

If your company and product line are significant enough, an external ecosystem will emerge to explore opportunities for collaboration with you. The creation of such an ecosystem offers large companies an advantage that smaller firms and startups lack. The team must exploit this by pursuing flow resonance across external ecosystems and within internal chain-link systems. This effort is continuous and vital for keeping pace with the Accelerating Change Race.

External Systems and The Accelerating Change Race

External systems are just as important as the product line system. Although they are separate from the product line, teams must devote focused attention to them.

Teams must develop the ability to learn from and engage with these system actors. A well-crafted enabling agent will facilitate this task. Chapter 20 provides more information on system enablers.

Chapter 14 Summary

External Systems and Performance

- **Internal and External Ecosystems:** Product lines rely on systems both within and outside the company. These systems are called external systems. Within a company, sales, finance, and operations are external systems that integrate with and support a product line system. Outside partners and suppliers make up other external systems. All external and internal systems, including the product line, comprise an entire ecosystem. Every product line system operates within a much larger ecosystem.

- **Internal Chain-Link Ecosystem**: Groups like sales and operations are "links" within the business strategy "chain." The product line is another link. Any weak link hurts the entire approach. Typically, no one oversees all the links and flows affecting a product line. People primarily focus on their own roles. However, product teams must intentionally align all internal groups and external partners. Proactive ecosystem alignment needs to foster the ability to improve the line's Outcome Velocity and execute strategic moves more quickly. Performing and transforming are far more challenging without ecosystem coordination and collaboration.

- **Aligning Flows:** Different people manage these external systems' workflows, cash, information, and decision-making flows. If those people don't coordinate, things won't line up between the groups. When systems don't match up, it slows down and undercuts the results.

- **Building Resonance:** All flows—work, decisions, data, and results—need alignment across external systems. Well-aligned flows create "resonance" that can boost performance beyond what any group could do alone. However, conflicts, with some parts focusing on efficiency and others on innovation, can cause problems.

Chapter 15

Focusing via Enabling Constraints

Strategy and Charters: Focusing and Guiding

He biggest challenge teams face in product line management is that they cannot influence or change every current and potential system component in ways that benefit the line. It's simply too much work and would require too much time. Even worse, much of this work would be unnecessary and waste resources. However, that does not mean we should give up. We can use a key system principle to help manage this complexity.

The 80/10 Rule

If you review the lessons taught by the systems thinking gurus mentioned in Chapter 5, you'll find several principles that help us understand complex systems. One is the Pareto principle, also known as the 80:20 rule. Generally, it suggests that 20% of the system's agents and forces account for 80% of its outcomes. However, after studying and helping develop many product lines,

I've learned that product line systems need to have a sharper focus than what the 80:20 rule indicates.

My experience suggests that stellar product lines have ratios closer to 80:10. This ratio shows that 10% or fewer of the system agents and forces generate 80% or more of a product line's results. This principle does not mean you can ignore the other 90% of the system agents. Instead, it redirects a product line team's focus toward certain parts, forces, constraints, actors, and enablers that are more important than others. The steeper ratio is highly helpful when dealing with uncertainty.

Teams gain more influence over their line when they recognize higher ratios. A strategy that intentionally concentrates on a top 10% is more effective than one that spreads resources too thin. It guides product line leadership to establish roles and assign responsibilities aimed at supporting and improving the top 10% of agents and forces. Teams can sharpen their focus using organizational structure and by redirecting agent enablers such as processes, methods, and practices—deliberate enablers—to support work, information, and decisions related to the critical parts and agents.

Strategy: The Important Enabling Constraint

Strategy directs a deliberate effort to improve a system's performance. It acts as an enabling constraint that establishes boundaries and guides the system's flow toward desired outcomes. Without this positive constraint, the system's flow and emergence would be unpredictable. You would have to rely on hope and luck to influence your line's performance.

It is important to understand that strategies are enabling constraints. But first, recognize that you are dealing with a complex system, not a simple one. Simple systems need controls, not enabling constraints. If you remove the main enabling constraint from a complex system, in our case, act without a strategy, the system will drift toward chaos and turbulence.

Consider that if a product line is an orderly or complicated system rather than a complex one, we should seek the root cause of what's hindering its performance. That focus, however, wouldn't be 80:10; instead, it would be closer to 100:1.

Product lines are complex systems shaped by various factors. There is no single root cause that limits performance. The way forward should be more targeted than random, but not overly broad. That's what the 80:10 rule helps teams focus on to enhance performance.

Tighter Focus

If a team finds it necessary to expand its focus beyond the 80:20 Pareto rule to include more system agents, it may need to reconsider, transform, or pivot its product line. The factors influencing this might be too scattered to allow the team to increase the line's Outcome Velocity. This suggests that the product line needs additional components and forces to move closer to an 80-10 ratio; therefore, the product line requires a new approach and strategy. Its structure must change, and sometimes, even the archetype must be altered.

It's important to consider how aging and maturity affect a product line. Teams often add more variations to their offerings as their product line system ages. When a line's platform-lever(s) reach a mature stage, you'll notice a team spreading its focus across many more system agents. Significant improvements in

Outcome Velocity become unlikely at a certain point in a platform-lever's life cycle.

Advancing the System

Changes in product line systems aren't just about reconfiguring or adjusting the components. A key characteristic of product line systems is that teams can modify entire categories or types of parts. Consider positioning moves. Product positioning across segments is often a crucial part of a line's strategy and requires deep insights into customers within each market segment.

The key part of the system that reveals customer needs is the Verb-based segment. These market groupings represent clusters of desired outcomes for the Jobs-to-be-Done (JTBD). This segmentation originates from the excellent work by Tony Ulwick and his team at Strategyn.[61] They called it "Outcome-Driven Innovation." Harvard Professor Clayton Christensen advanced the concept with his book, "Competing Against Luck."[62] Christensen refers to Ulwick's outcomes as jobs that customers want done.

I refer to market-defined clusters of the jobs "Verb-based segments," i.e., segmenting a market based on Ulwick's outcomes or Christensen's Job-to-be-Done. I also include Verb-based segments as key parts in a product line stack.

Jobs-to-be-Done theory focuses on customer satisfaction. The theory is based on the idea that people "hire" products to complete a "Job." JTBD practices help us assess how satisfied a current or potential customer is by accomplishing a "job" with a product. A gap in customer satisfaction, meaning a job not being done, creates an opportunity for a new product. JTBD also points out where we might gain an advantage over competitors because they cannot fulfill certain jobs.

Using Verb-based segments greatly benefits a product line. Before creating a product, these system elements assist teams in focusing on two key factors that drive the line's progress— customer satisfaction and competitiveness. Not surprisingly, Professor Christensen mentioned that the JTBD theory is far more effective than relying on luck and hope, which we often turn to for innovation decisions. Appendix 7 provides additional details on verb-based segments.

Boundary Defining Charters

In product line systems, JTBD unfolds through two interconnected components: the Verb-based market segment and the product innovation charter. The latter, the innovation charter, acts as an enabling constraint for creating products, platform-levers, and technologies.

I credit Professor Merle Crawford for introducing us to innovation charters. I know this because he was both a friend and a mentor to me. His research guided us in setting targets for innovation, which he called product innovation charters.[63]

Interestingly, Crawford concentrated his innovation charter efforts in the 1980s on supporting teams. During that time, cross-functional teaming and team charters were popular academic topics.

Today, with a deeper understanding of complex systems, we see how they serve as enabling constraints for development projects. In Appendix 7, I detail the process of creating and using innovation charters. I also explore how JTBD research influences verb-based market segments and shapes outcomes embedded in charters. In Chapter 29, I also demonstrate how innovation charters act as a key element for leveraging AI.

Agents and Their Interplay

We recognize that the interplay among system agents is essential because it influences the performance and adaptability of the product line. Participating in the Accelerating Change Race begins with exploring and experimenting with the dynamics of a line's key parts and agents. The decisions and choices made by individuals, along with insights gained from experimenting at the intersection of components and agents, direct a product line toward efficiency or change.

There's little doubt that a genius among us will discover a groundbreaking platform-lever and system archetype that will boost product line velocities beyond anyone's expectations. Meanwhile, most companies must dig deep and figure out how to keep their lines relevant.

Free of Constraints

Many people prefer their activities to be free of limits or boundaries. Lean and Theory of Constraints practitioners avoid embracing constraints or limitations. These experts focus on avoiding, removing, or reducing constraints. Creative workers who seek or create new opportunities share similar views. They see constraints as the box outside of which they need to think.

An Enabling Constraint serves a different purpose in complex systems. It establishes boundaries; or, put another way, it provides focus and guidance that help teams achieve better results more quickly. Enabling Constraints define boundaries that guide judgments, work, and information.

However, remember that completing projects quickly is not the main goal. Instead, teams must focus on improving cash flow, customer satisfaction, and competitive positioning. The most important goal is to boost the line's outcome speed both now and in the future.

The Strategy Questions: Where to Play and How to Play

Roger Martin, the renowned strategist, demonstrated that the essence of business strategy lies in answering the 'where to play' and 'how to win' questions. In his 2002 best-selling book "Playing to Win: How Strategy Really Works"[64], Martin eloquently argues that the key to success is addressing these questions clearly and enabling your organization to establish and implement work and decisions coherently and cohesively.

Answers to the Where and How to Play questions are crucial for developing a strategy that improves business performance. The responses need to align with insights into product lines and the different approaches companies can take toward customers, products, and technologies. These responses should be integrated into enabling constraints.

Product Line Indivisibility

Product offerings and innovation remain central to the business's "Where-and-How" answers. The indivisibility of a product line strategy (Chapter 2) makes it a key to uncovering those answers.

A company's strategy—its main guiding constraint—sets boundaries that promote coherence and a sense of unified diversity within the system. These boundaries allow contributors to recognize, evaluate, and select beneficial actions and disregard those that are not. They help steer focus, enabling teams to concentrate on the most achievable yet impactful nearby opportunities.

Understanding Noun-based segments offers valuable insights into the Where-to-Play question. Meanwhile, the results outlined in Verb-based segments set targets for developing and delivering attributes. A thorough understanding of these targets and how to deliver attributes that match them defines the playing field for addressing the How-to-Play question.

The How-to-Play answer outlines the company's strategy for developing and delivering attribute sets that align with the verb-based outcome clusters. It should clarify how the system maximizes and continuously improves the line's Outcome Velocity, thereby enhancing customer satisfaction, boosting cash flow, and maintaining strong competitiveness.

This concept brings us to a fundamental PLSM principle. Teams must identify the Where (industries and target noun-based market segments) and the How (using platform-levers, technology building blocks, enablers, and ecosystems) by intentionally focusing on a specific 10% of the system's agents. This focus shapes the product line strategy—the flexible boundaries set as the system's main enabling constraint.

Product Line Strategy

In my 2018 book, The Profound Impact of Product Line Strategy[65], I discussed three components of forming a product line strategy, which I referred to as the "Strategy-Essence." These components include Platform-leverage, Product Positioning, and non-product Chain-link Alignment, all of which are essential to the performance of a complex product line system.

As companies improve their PLSM capabilities, product line teams must further define the system agents related to the Strategy Essence. Platform-levers must clarify the system's archetype and describe the intensity of focus required for each one. Teams must also clearly articulate product positioning for each Verb-based segment and relative to each other. Additionally, teams need to manage and influence Chain-Link alignment while pursuing synergies across the links.

Proximate Goals

Many managers get stuck when developing product line strategies and creating project charters because they try to define specific goals and objectives. This approach is not recommended for complex dispositional systems, such as product lines. Instead, the goals and objectives should be "proximate," not exact. Focusing on specific goals and objectives works well for orderly, not complex systems.

Product line teams should avoid being overly focused on specific objectives. System goals and targets with fixed end dates can be counterproductive. They redirect focus to immediate concerns, which can hinder agility.

The major challenge is overcoming the strict adherence to the orderly system practices of old-school managers. Unfortunately, complexity theory has yet to permeate every manager's thinking.

Constraints that Help

An enabling constraint doesn't obstruct a system's progress. Instead, it establishes boundaries that direct and optimize the system's flows.

Enabling Constraints and The Accelerating Change Race

A "constraint" often has a negative connotation in project management, suggesting something to avoid or escape from. However, complexity theory highlights the advantages of "Enabling Constraints," which establish boundaries rather than barriers. These constraints help focus work and guide decision-making. An enabling Constraint is especially useful when uncertainty is high.

I identify two types of Enabling Constraints in product line management. One defines strategy, while the other establishes a charter to guide a project. A strategy acts as an enabling constraint that directs the entire product line and focuses on a relatively small number of system components or agents.

Charters define boundaries for creating and developing strategic moves, parts, and agents. The most common charter is for developing new products, known as a "Product Innovation Charter." A charter provides guidelines that ensure coherence between projects and the product line.

Without enabling constraints, it would be nearly impossible to keep pace in the Accelerating Change Race.

Chapter 15 Summary

Enabling Constraints: Strategy and Charters

- **Challenge of System Forces: It** is often difficult to ascertain how each force affects the product line Outcome Velocity. The forces interact and influence each other and the system's parts.

- **The 90-10 Rule and Enabling Constraints:** **Unlike** the Pareto principle (80-20 rule), most product lines need a sharper ratio. A critical 10% of parts and forces will drive 90% of a product line's outcomes. This system's principle presents as an Enabling Constraint—the strategy—that guides teams and aids in prioritizing essential actions to achieve efficiency and change.

- **Major Product Line System Advancements:** Changes in a product line system involve advancements in entire categories or parts and forces. Verb-based segments and innovation charters are crucial to improving a product line's Outcome Velocity.

- **Strategy Questions - Where to Play and How to Play:** Roger Martin's strategy questions[61], "Where to play" and "How to win," are foundational for superior business performance. For product lines, understanding Noun-based and Verb-based segments guides these answers, emphasizing the importance of focusing on a select 10% of critical parts to define the product line strategy.

Chapter 16

Strategic Moves & Transformative Shifts

Growth by Advancing, Extending, Pivoting, and Transforming a Product Line

Strategic Moves are accumulations of "adjacent possibles" for the product line. They build on each other to move the line from its current state to a new one. We consider Strategic Moves as "adjacent possibles" because they are the simplest next steps for expanding the line. We describe the line's progress in terms of customer satisfaction, competitiveness, and cash flow. When well-coordinated, one strategic move can open up opportunities for others. The result is the compounding of gains, not just their sum.

Transformative Shifts represent a higher order of strategic moves. They involve a change in the system archetype. These strategic moves purposefully alter a product line's architecture and ecosystem, leading to significantly different outcomes that promote growth. Several strategic moves may be needed during and after a transformative shift.

E very product line team's mission is to continuously improve its Outcome Velocity in pursuit of growth. The goal isn't just to hit a target and call it done. Instead, it's about creating a series of projects and initiatives that can positively impact the line's Outcome Velocity both in the short and long term. These actions are the product line team's "strategic moves" that put the strategy into action to grow the line.

Strategy and Its Execution

Some managers believe that strategy and execution should be separate. However, this view isn't helpful in product line management. I like to use an analogy of radio waves to show why they need to work together.

Just as electrical and magnetic waves intertwine to produce a radio wave, the workflow and decision flows of the product line system interconnect and are shaped (modulated) by the information flow. Together, these flows generate the outcome flow, which represents the product line system's radio wave.

Strategy serves as the waveguide for the product line's execution. Anyone who has worked with waveguides knows how sensitive these things can be. Major performance issues arise if the waveguide doesn't match the radio wave.

In our analogy, problems occur when the product line strategy doesn't match the system flows. The strategy (the waveguide) or the workflow and decision process (the electromagnetic wave) may need adjustments. The key is the alignment. Ideally, teams adjust the workflow and decision-making processes to fit the strategy. However, teams might also need to change the strategy to match what is feasible within those flows.

Coherent, Compounding Moves

Product line teams must answer two questions to develop effective strategic moves. First, how should they create and execute each move? Second, how do they ensure these moves produce a continuous stream where each one increases the value of the others? This stream should aim to boost the line's Outcome Velocity and provide more value and growth than individual moves could achieve.

To answer these questions, we must step back and recognize that we are working with a complex system. The solutions should align with our understanding of complexity theory.

The most important insight complexity theory offers about this challenge is that a system's performance relies on how its components interact. To improve the system, you need to enhance the interactions among its current and potential elements and agents. Agent Interplay and The Adjacent Possibles

System actors, especially team members, understand the interactions among system components, forces, and enablers. Those who are well-versed in the product line system can often determine whether a change in a part or agent will enhance the line's Outcome Velocity. This knowledge and awareness also make these individuals the most effective contributors when exploring the relationships among the key parts and agents.

The typical approach to creating products involves identifying customer needs and exploring new features to address those needs. The goal is to develop products with distinctive features that attract customers. But, this traditional approach depends on a single-product mentality.

Every team's goal is to enhance its line's Outcome Velocity in both the short term and long term. This requires making numerous judgments and decisions about key system components, enablers, and insights into what is beneficial and what is not. These decisions are based on their understanding of the system, its constraints, and the forces at play. Such judgments guide further exploration and insight-gathering. Ultimately, the team needs to combine insights and knowledge to develop strategic actions. Then, they must evaluate the effectiveness of each move within the entire set of strategy moves.

On The Roadmap

A product line team must aim to develop a series of actions that multiply (think: increase exponentially, not just add) the Outcome Velocity gains. This task is more difficult than creating one isolated action at a time. The actions need to connect to and benefit from the system's other components. I will discuss this topic in upcoming chapters as I introduce the use of innovation charters and sensemaking.

The strategy, the system's main enabling constraint, should direct the flow of moves. Each move must serve a clear purpose and intent, clarified through charters and other guiding constraints. Charters set boundaries and help focus work efforts. However, it's not just the natural interaction between agents that creates meaningful strategic moves. You'll find that the job demands a certain level of intelligence.

Insight Affordance

The most important aspect of exploring agent interaction involves establishing an environment and providing the tools (enablers) that support this work. I refer to the environment and tools as insight affordance. Chapter 18 is dedicated to creating insight-generating affordance.

Affordance is an invitation to think and act. It combines an environment, resources, and encouragement connected to a sense of purpose. It does not involve telling people what work to do or how to do it. Instead, it invites individual contributors, whether working alone or in groups, to do what they believe is best. That, for many, is a radical concept.

Weed-and-Feed for Outcome Velocity Gains

Encouraging and managing insight discovery is key to improving a product line's Outcome Velocity. However, the goal isn't to rush to generate insights faster. Instead, I view the process more as a managerial approach, akin to weed-and-feed lawn care. First, we weed out insights unlikely to improve the line's Outcome Velocity. Then, we nurture insights that show promise for boosting the line's Outcome Velocity. We provide this nurturing by allocating more resources to them.

Many insight discovery venues and events should occur before a project, or a multi-project initiative, starts. These efforts don't need to be planned or assigned in advance. It indicates a highly effective group when individual contributors take the initiative to start such efforts independently. It's an even stronger sign when self-formed leaders recruit contributors to join the work. The topic of distributed leadership is so crucial that I dedicate Chapter 19 to it.

Ultimately, teams need insights to make meaningful decisions. However, moving from having insight to making a strategic move requires collective intelligence. I can't offer a magic formula for developing that intelligence. It requires strategic, creative, analytical, and systems thinking working together. It also needs diverse perspectives and experiences, often shared through stories and analogies. Still, I suggest that teams become familiar with different possibilities and potential strategic moves, regardless of whether they seem applicable. At some stage, participants should design and evaluate these potential moves. I list possible strategic moves in Table 16.1 below.

Table 16.1 Potential strategic moves all product line teams should consider.

Move Description	Strategic Move Type	Effects on the system's Archetype	Potential Impact on Outcome Velocity	Needed Chain-link Changes
Developing and launching a **New Product** (PIC → PID → PIM) progression	Advancement	None	Incremental	None to little
Radical New Product based on major new technology building block (without changing platform-levers)	Advancement	None	Some to Strong	Some likely
Improving the effectiveness or efficiency of a **Platform-lever** without changing its platform-lever type; Platform-lever multi-generational planning; a new generation of PIMs	Advancement	None	Some	Some

Move Description	Strategic Move Type	Effects on the system's Archetype	Potential Impact on Outcome Velocity	Needed Chain-link Changes
Changing the **Platform-lever Mix** without changing the system archetype, upgrading one or more platform-levers simultaneously.	Advancement	Sometimes, when the platform-lever mix changes significantly	Some	Some
Pursuing **Adjacent Markets**	Advancement	None	Some	Some likey
Shifting the line's **Primary Marketing Focus** to new markets or market segments	Pivot	None	Strong	Some to Notable
Re-segmenting the Market to recast product offerings toward new/different customer needs	Pivot	None	Strong	Some to Notable
Changing the product line's **Archetype**	Transformation	Major	Major	Major-Significant
Adding **New Technology Building Blocks** to enable new products to gain new features and attributes.	Advancement	None	Some to Strong	Some
Cost Reduction	Advancement	None	Some	Some likely

221

Move Description	Strategic Move Type	Effects on the system's Archetype	Potential Impact on Outcome Velocity	Needed Chain-link Changes
Resiliency and Change Capacity moves	Advancement	None, except when major	None, except when a change is required	None, except when a change is required

Strategy execution involves a wide range of actions. However, many of these actions are not related to product development; they are not intended to produce products. Instead, they focus on creating or modifying all system components and agents, as well as their interactions.

The coherence and relationships among these moves are crucial for improving the line's Outcome Velocity. Teams maintain coherence and coordinate their strategic moves through the Roadmapping process. In Chapter 21, I discuss this process and the practices that support it.

Let's first look at a few basic strategy execution moves.

- Advancing The Strategy: "How to Play"

Innovation Charters are key parts of product line systems. I cover and explain them in Chapter 15 and Appendix 7. A charter acts as a target that sets boundaries for developing projects. It makes sure the project aligns with the product line system and enhances Outcome Velocity. However, when we use Innovation Charters, we also need to understand how a charter flows and changes within the system. Knowing this flow can be especially helpful for many managers because the Innovation Charter is a new idea for them.

Teams document the key elements of an Innovation Charter. This document acts as a guiding constraint for idea development; it neither predicts nor limits the outcomes. In PLSM, charters specify and direct development projects, which can involve various system components such as new products, platform-levers, composition blocks, and technology building blocks. Teams may also use charters to scope and define Verb- and Noun-based market segment research.

The innovation charter sets guidelines for possible work and decisions. For example, when the initiative centers on a new product, the charter specifies customer-focused Jobs-to-be-Done outcomes for a Verb-based segment. It also reflects the product's purpose related to the system's performance and the line's strategy. It highlights expected improvements in outcome speed and details possible relationships with other system components.

Developing, refining, and investing in an innovation charter is a deliberate process. Once an innovation charter is chosen for investment, it is incorporated into the product line roadmap (for products) or the assemblage map (for agents). Its position on the map should indicate the expected completion date and its relationship to the platform-levers and other system elements, such as technologies and market segments.

Many companies have set up portfolio management specifically for new product development projects. In product line systems management, we expand portfolio management to include work on innovation charters. The key resource management question becomes whether resource allocation across the entire portfolio aligns with the product line roadmap. Note that these system enablers (portfolio management and roadmapping) must be synchronized.

- Pivots: Redirecting "Where-to-Play"

Recognize that transformations are different from pivots. Pivots are simpler than transformations. A pivot keeps the core concept while shifting to different market segments. In such cases, a team may carry out several market or segment-specific charters that set boundaries for segment analysis work, aiming to guide or influence a deliberate pivot.

Most likely, pivots will involve new product development projects (PIDs), each aimed at delivering a new attribute set. These appear as Product Innovation Charters (PICs). However, pivots don't require new product line strategies or changes to business models. While sales and marketing need to realign with the new segments, their work processes will stay the same.

- Transformative Shifts: Changing Archetypes

The most significant strategic move occurs when a team changes the archetype of its product line. This shift transforms the product line and requires a deep understanding of it, as well as a nuanced grasp of the system, functional chain-links, and ecosystem contributors. Also, an archetype change demands that the strategy of the line also changes. Furthermore, supporting organizational structure, processes, practices, and IT system enablers must also be adjusted accordingly.

Pursuing Growth

In this book, I explore how product line systems are built and managed. Now, I want to focus on the main goal of managing these systems, which I can sum up in one word: **Growth.**

Product line teams must prioritize both efficiency and change to ensure the line can survive and thrive. Survival alone, however, is not enough; growth is crucial. Business growth is similar to happiness in human experiences. Without growth in a product line, the complex system is likely to collapse if a team focuses only on survival.

The good news is that growth is in the organization's control. External forces don't solely determine growth, as some might believe. There are always chances for growth. The key question is whether you can improve your line's efficiency and make the necessary changes to succeed. The answer depends on the strategic actions a team takes and executes. These actions are vital in deciding whether a line will grow, stall, or shrink.

Identifying the most effective strategic moves, or at least the preferable ones, is crucial for achieving growth. The need for growth emphasizes the importance of adjusting and coordinating the components, enablers, participants, and architecture of a product line system.

Strategic moves must coordinate effectively and align with the system's components and enablers. The key is to embed the pursuit of growth into the line's strategy, one that guides a team's focus and the line's direction. This advice aligns with that given by strategist Roger Martin.[67] Below, I've translated Martin's approach into managing product lines as systems.

The following three growth tactics should guide product line systems management. I present them in order of importance. They are not intended to be sequential. You do not need to complete one before starting the next.

Seek Strategic Moves To:

1. **Increase your market share in your Noun-based segment.**

- Focus on revenue share, not units sold.

2. Grow the size of each Noun-based market segment

-   Create   offerings   to   upsell   products.
- Attract new customers to each noun base segment through targeted JTBD-driven (Verb-based) product attribute positioning. Utilize both engineered-tangible positioning and marketing-perceptual positioning.

3. Pursue adjacent market opportunities

- Identify adjacent (Noun and Verb-based) markets where altering your line can deliver products that satisfy customers, become competitive, and drive greater cash flow.

Throughout your efforts, teams need to optimize their investment of money and resources across all projects as a unified whole, rather than focusing only on individual projects. This involves connecting the project portfolio to the product line roadmap. The goal is to shift investments to support desired growth outcomes and to create coherence by considering the preferred portfolio mix over different time horizons. The product line roadmap helps align the project mix with current and future portfolios. If they don't align, then either the portfolio mix or the roadmap and strategy must be adjusted.

Perform and Transform

We've all heard stories or read articles about digital transformations and organizational restructuring. Many of us have contributed to these change efforts. However, here's a warning: do not confuse digital transformations with product line transformations. They are different.

Most digital transformations focus on improving efficiency in product offerings. They aim to change people, organizations, workflows, and IT systems rather than modifying the products or product lines themselves. When a large company revamps its product offerings, it's usually because the firm faces significant threats. I mention this not to judge whether it's good or bad, but to show that it's a common occurrence.

In recent years, the leadership mantra, "Perform and Transform," has become firmly established in many C-suites. It represents the organization-wide equivalent of addressing the tug-of-war between product efficiency and change. A deeper dive into the Perform and Transform movement reveals it as a challenge related to leadership, culture, and mindset (people's behaviors, attitudes, and understanding) linked to digital transformation.

Today, digital transformation efforts often overlook simultaneous efficiency and change in product innovation and management. This needs to change. Integrated data systems enhance product flow and key part creation. However, building or transforming these data systems must happen within the context of a multi-product, complex system.

Strategic Moves, Transformations, and Accelerating Change Race

Product line teams must enhance their line's Outcome Velocity by creating and nurturing a steady flow of "Strategic Moves." To form and implement such moves, teams need to understand and adapt to the dynamics among the agents in the complex product line system.

Teams should develop and implement strategic moves to remain competitive in the Accelerating Change Race. If necessary, these moves must alter the line and its architecture (i.e., change the system archetype and all associated parts and agents).

Chapter 16 Summary

Strategic Moves and Transformative Shifts

- **Product Line Mission:** Each product line team aims to enhance its Outcome Velocity continuously, not just achieve a goal and then stop. This involves creating a continuous stream of projects and initiatives.

- **Strategic Moves:** The team's actions and initiatives that positively affect the product line's short- and long-term Outcome Velocity.

- **Transformative Shifts:** This major strategic move seeks to change the line's archetype, necessitating significant changes to the line's key parts and agents.

- **Complex System Approach:** By acknowledging that product lines are complex systems, we recognize that a product line's improved performance depends on the interplay among its key parts and agents.

- **Insights and Affordance:** Valuable strategic moves are built from insights gained through exploring and experimenting with the product line's current and potential agents. Creating an environment that supports insight generation, referred to as "affordance," is crucial.

- **Roadmap and Coherence:** Strategic moves must align and complement each other. The roadmap process ensures coherence and orchestrates the team's moves to advance the product line's Outcome Velocity.

- **Strategic Move Examples:** There are many types of strategic moves. Key moves include product developments, platform advancements, pivots, and archetype transformations. Each move should relate to the others.

SECTION FOUR

Influencing Outcomes

Insights | People | Structure

"You may hate gravity, but gravity doesn't care."
– Clayton M. Christensen; Professor, Harvard Business School

Product line management involves influencing and nudging the emergence of the system's outcome. When you do that better than your competitors, you win.

- The Author

Chapter 17

Insights, Charters, & Sensemaking

Assumptions, Heuristics, and Safe-to-Fail Experiments

In the last chapter, I discussed the importance of generating insights and using them to create a series of strategic moves. This may be the most essential skill a product line team can acquire.

Insights represent new pieces of knowledge and understanding. Learning occurs when insights are thoughtfully combined with existing knowledge. In product line management, accumulated learning decreases uncertainty and informs strategic decisions.

All strategic moves begin with insights considered "sensible" and helpful regarding the product line system and its performance. Teams must strive to enhance insight generation and product line sensemaking, leading to impactful strategic moves. The key is to concentrate insight-generating efforts on the interactions among the system's current and potential components and agents.

Many managers assert they know how to generate insights, and for the most part, they are right. The problem is that they perceive this as a creative task intended for a select few. They also believe this work should concentrate on creating and improving a single product. However, for product lines, it is vital to generate insights about the different components of the system, their interactions, and the behavior of the overall system. Everyone involved in the line, not just a few, should be part of this effort.

This chapter examines insight generation, sensemaking, and the formulation of strategic moves to maintain pace in the Accelerating Change Race.

The Insight-Sensibility Flow

Insights serve as nutrients for product line systems. They enhance information flow, guide decision-making, and provide purpose to the workflow. They serve as fodder for formulating strategic moves. Greg Githens, a leading expert on Strategic Thinking, calls insights the "magic sauce" of strategy.[67]

Product line teams must learn to generate meaningful insights and combine them with existing knowledge to create a series of strategic moves that build upon one another. Although people often aim to link insight generation exclusively to product innovation, this is not the only application of insights that can improve a line's Outcome Velocity. Other system components, not just products, also play a vital role in making impactful moves. Nevertheless, all strategic decisions start with insights. If teams fail to generate meaningful insights, their future will be bleak.

Big Companies, Unique Challenges

Entrepreneurs are fortunate because they usually engage with only a few ideas from a select group of individuals. However, they also face a disadvantage since they have a limited number of people to evaluate those ideas.

Large companies with established product lines confront a distinctly different challenge. They have many individuals who possess diverse knowledge, and their problem lies in making sense of the ideas and insights that everyone has to offer.

Consider the number of people involved and the range of innovative ideas Apple must evaluate to maintain a consistent flow of strategic actions driving its iPhone product line. Or think about the volume and importance of insights Ford needs to develop to establish a strong foothold in the EV market. No individual, not even Steve Jobs or Elon Musk, can generate and transform so many insights into a cohesive stream of strategic moves. There must be another way.

In large companies, the challenge affects both individuals and the entire organization. Individuals need to generate insights, while the company must facilitate the integration of these insights into their work and decision-making processes. This challenge is both local and personal, as well as system-wide and organizational.

Focusing on Insight Generation

Two focal points for generating insights are most effective in leading to strategic moves that drive both near- and long-term gains in product line Outcome Velocity. The first focal point is **the interactions of current and potential system components and agents.** The second focuses on **the system's architecture regarding constraints and forces.**

You can facilitate organizational change and instill attitudinal "mindsets" as much as you want. However, your line will inevitably decline if you cannot deliver meaningful insights about market segments, platform-levers, or other system components and agents. Such insights are the bread and butter of product line management. Nevertheless, the importance of insight lies in sensibility, not just in the ability to use it to enhance the line's performance.

For example, if an insight helps lower costs and drive efficiency, it may seem attractive. However, its implementation might also create organizational inertia, making change more difficult. The question is whether pursuing this cost-cutting strategy is sensible. We can only evaluate that by examining the move's probable effects on the system in both the near and long term. Insight sensemaking is not merely a feasibility test; it is also a strategic inquiry.

- Parts and Agents

In complex systems management, the sensibility challenge can refocus efforts to generate insights. It begins by directing work towards the interactions among the system's components and agents. Subsequently, knowledge and insight evolve into sensemaking that is specific to the line and system. Learning can then redirect insight generation towards other existing and potential components and agents. It's an iterative activity.

A key challenge for many companies is to keep the right parts available and ready when needed. They must find new parts, develop their readiness, and understand the necessary system changes to support their contributions. To achieve this, teams need to make sense of many moving components, which requires insights. However, many companies delay starting this work until they feel the pain of failing to change. Such delays always result in late strategic moves and numerous cascading consequences. (See Chapter 11, Urgency before Emergency)

- The System Architecture (and Archetype)

A product line architecture significantly influences the behavior of the system and how the components and agents of a product line collaborate. It also impacts the line's ability to adapt. Many of us have observed numerous examples of this in software product lines over the past decade or two, as various software companies modified their products to operate online as a service. Customers sought this change to enable software use anytime and anywhere. To achieve this, companies needed to gain deep insights into their current and future platform-levers and system archetypes.

The Software-as-a-Service (SaaS) transformation required each company to generate new insights based on emerging skills, programming languages, operations, and security approaches. The job involved more than simply "re-platforming" the product line. Each product line team needed relevant insights that aligned with their newly transformed system.

The challenge of forming meaningful insights also applies to non-software platform-levers such as design, production, or service platform-levers. This challenge can be confusing because all parts will age. Maturity puts the line at risk, and in response, teams must change these platform-levers, not just improve the surrounding parts and agents. These ongoing challenges emphasize the need for a continuous flow of meaningful insights.

Sensemaking and Charters

Understanding is fundamental to human existence. Unsurprisingly, it is also central to business management and essential for managing product lines. However, each domain differs based on its constraints, forces, and system components.

The business management community was introduced to "Sensemaking" by University of Michigan professor Karl Weick in his 1995 seminal publication, Sensemaking in Organizations.[68] Sensemaking is the broader task of stitching together insights with existing knowledge to create understanding within a context. For us, that context is the complex product line system.

A helpful framework for sensemaking is the OODA Loop, which stands for Observe, Orient, Decide, and Act.[69,70] It is particularly valuable in complex systems because it enables individuals to understand situations by cycling through the loop. Military strategist and US Air Force fighter pilot Col. John Boyd developed OODA during his leadership role in the Vietnam War. Rapid iterations through the OODA loop have proven effective in highly uncertain environments, especially when the outcome, i.e., the act, may involve experiments to gain further insights.

PLSM provides three key contributions to improve product innovation and management sensemaking.

The first contribution is the shift in context to a multi-product, complex system perspective. The PLSM context shift alters what some might describe as your "mindset" or what you think about. If people understand and embrace the context shift, their "mindsets" will follow. PLSM does not advocate for changing mindsets; however, it does require a change in context. This is crucial because sensemaking is highly context-specific. We need insights to make sense in a complex, multi-product system environment.

The second contribution involves adding innovation charters to the system, which creates a pathway for knowledge and insights that enhance it. The charter is a conceptual object representing a potential strategic move, connecting to other parts and agents within the system when placed on the roadmap model. It is these connections and their potential influences on the system that guide sensemaking. A team must question whether a charter, along with its embodiment of insights, may lead to a strategic move that benefits the system. If the team believes it does, then it is sensible. Each charter also serves as an "enabling constraint" that establishes a focus for teams to develop a system component or strategic move. See Appendix 7 for more on charters.

The third contribution is the intentional and purposeful encouragement of all system contributors and actors to generate insights and engage in product line-specific sensemaking. System experts refer to this encouragement as the line's "affordance." It involves creating the environment, tools, and motivations necessary to pursue insights about the line's components and agents and to participate in meaningful sensemaking.

The overarching principle of these contributors is to ensure the constant, deliberate, and open pursuit of insights and knowledge to improve a line's Outcome Velocity and greater capacity for change. This pursuit differs from traditional single-product management, which focuses on discovering new product concepts and enhancing the efficiency of existing products. The PLSM approach enables individual contributors and ad hoc teams to delve deeper into insight generation and product-line-specific sensemaking.

Sensemaking and Uncertainty Dampening

Generating insights without sensemaking is a futile pursuit. If teams aim to produce insights to address the efficiency-change dilemma, they must combine insight generation with system-specific sensemaking.

The interaction between insight generation and sense-making is neither linear nor static. When managers examine how insights might affect a product line system, they are likely to refine the insights and integrate them with other knowledge. This process creates a learning loop of discovery and assessment. From this learning loop, innovation charters for potential strategic moves emerge. Notably, this effort also enhances potential gains and reduces the uncertainty surrounding those moves.

I view the insight-sensemaking learning loop as a powerful vortex that draws in knowledge and other insights while filtering out what is unhelpful. This dynamic vortex occurs only when managers and contributors have a deep understanding of their product line. The challenge is to initiate the vortex and increase its rotational speed. I address this in the next chapter on PLSM affordance.

Sensemaking and Organizational Learning

Events, discoveries, and changes in the world can lead to insights that impact a product line. Some are obvious or well-known, affecting your product line just as they impact your competitor's line. These events are newsworthy, appearing in many sources, and resonating as loud signals above the daily noise.

Many journalists and industry experts earn their living by conveying strong signals. They also communicate the actions taken in response to these signals. All teams can easily access this information. Consider the numerous stories we read today about artificial intelligence or the economy. These serve as strong signals, although they can be debated endlessly.

The greater challenge for product line teams is to identify and interpret weak signals before they develop into stronger ones. In a competitive environment, the goal is to derive insights from these faint signals and transform them into effective actions before competitors do. However, not all companies notice every weak signal, nor do all departments perceive weak signals in the same way.

Weak Signal Sensing

Product line teams must recognize that their competitors are also searching for insights and trying to understand things. If a team can achieve this more effectively than its competitors, it can secure a significant advantage in making impactful strategic moves.

This job is similar to radio signal processing. I realized this through a hobby I engaged in as a young American boy living in Germany during the early 1960s, feeling isolated from the world. I was there because my father was an Army officer whose duties began with the Berlin crisis. I spoke no German and had no TV. The local English-speaking radio station was the Armed Forces Network, which aired repeat broadcasts of 1930s and 40s radio shows. My joy came from reconnecting with America through a shortwave radio. This passion has stayed with me since then.

Today, I hold an extra class ham radio license (W4NPD) and have extensively experimented with weak signal processing in both analog and digital waveforms. I follow the work of Princeton Physics professor Joe Taylor (K1JT) on this topic. The basic technique we use is similar to that employed in noise-cancelling headsets. It involves taking the "strong signal" and general noise from frequencies adjacent to where we are listening and inverting their waveforms 180 degrees out of phase. We then mix this opposite signal back into what we're listening to. This applies to both radio and audio waves. The negative phase addition neutralizes the strong signal—not perfectly, but significantly. It then reveals the weak signal for easier identification.

We can use the same approach to identify weak signals in information and data related to product lines. I've heard that government intelligence agencies do the same. They sift through all target and adjacent domains of information to extract the strong common signals that are visible to everyone, including competitors. They eliminate these strong signals from the source. They also apply this method to low noise-level signals (like static occurring on all frequencies). What remains is a collection of potential weak signals. The challenge, then, is to determine which signals are significant and which are not. This is where intelligence, wisdom, and intuition come into play.

Narratives and Sensemaking

Another source of insights and sensemaking comes from the stories people share. Such narratives can be extremely helpful because they often contain valuable insights paired with reasoning as a form of sensemaking. Look at any of Malcolm Gladwell's books to see this in its most direct form.[71] Whether you agree with the "sense" he lays out in each vignette is beside the point. (I don't, for many) Rather, the interesting part is his skillful use of narrative to communicate his sensemaking of a complex situation.

Most people glance over narratives without motivation to analyze them. That can be a huge mistake. Narratives are packaged communications that can be deconstructed into insights and meaning.

Generating Meaningful Insights

Generating insights that could influence the product line is easier when those producing insights have a strong understanding of the product line system. However, making strategic decisions based on these insights often requires exceptional, in-depth knowledge of the system.

Contributors who use creativity to explore topics and innovate products often express that they do not want to be restricted by constraints. Unfortunately, the lack of constraints or guidelines is why most insights, although interesting, remain ineffective for the product line system.

The job is not to create offerings. Instead, it is to generate insights that impact the line's performance. The team then uses these insights to formulate and reformulate strategic moves. However, keep in mind that teams need the moves to be coherent with one another and to address both sides of the new efficiency-change challenge.

Shortcuts in Complexity

Insights also arise from rearranging and testing the shortcuts we use to navigate and shape our understanding of our intricate product line system.

Research shows that humans consistently create shortcuts to guide us and expedite outcomes when working with complex systems. These shortcuts take the form of "assumptions" and rules of thumb called "heuristics." They develop from our experiences and learning, becoming embedded in our thinking. We tend to use them reflexively more often than intentionally.

These shortcuts may seem straightforward, but there is much to explore regarding assumptions and heuristics. Our main concern is how they can obstruct the development of insights.

- Assumptions

Assumptions are beliefs accepted without evidence. They can be helpful as they simplify complex situations and aid in dealing with uncertainty. For example, if you believe customers appreciate your brand, you might consider organizing a product launch with extensive advertising. Alternatively, your team might assume that a competitor will always respond to new technologies by lowering prices. This could lead the team to adjust how they position a product, a strategy they consider best for the line.

However, assumptions can also be problematic when they lead teams to make incorrect decisions or overlook crucial information. For instance, if your team believes customers love your brand, they might miss the fact that customers dislike certain product features. Similarly, if you assume your suppliers are reliable and that your supply chain operates smoothly, you might be taken aback when the quality of certain product components causes your product to fail.

It's important to be aware of assumptions and check them against reality. Teams can ask questions, seek evidence, or test each assumption. This approach helps teams avoid unintentionally harming their line. Moreover, it enables teams to gain new knowledge and develop insights that propel the line.

- Heuristics

Heuristics are rules of thumb that help teams avoid extensive analyses on various topics. They enable managers to make decisions without having to relearn the system and its components. These shortcuts allow teams to make quicker decisions and achieve results sooner.

Consider how doctors employ heuristics. A doctor's education shapes their heuristics. These diagnostic shortcuts enable doctors to quickly assess symptoms and determine the best course of action. Without these rapid rules of thumb, an emergency room would descend into chaos.

Teams need to understand their heuristics, yet most managers don't spend time examining them. They continue to rely on outdated shortcuts, despite changes in the product line. It's like a doctor from two hundred years ago attempting to run a modern hospital. There would be a significant difference between their diagnoses and the outcomes expected by the patient.

Curious Learning & Safe-to-Fail Experiments

There is a principle involved in reviewing heuristics: *everyone working on a product line system must learn, and they must do so continuously*. New knowledge and insights are essential to the system's lifeblood.

Team members should be curious and actively explore new insights into the system's behaviors, components, and architecture. They should question, investigate, and maintain a strong interest in the entire product line.

Experimentation has proven to be an effective way to generate insights, regardless of the context. However, we must recognize that some experiments may harm the line if the results aren't favorable. This is a much bigger deal at the product line level than for individual products, which underscores the importance of "Safe-to-Fail" experiments related to potential strategic moves.[72]

As a quick aside, one major challenge in generating insights is the inherent cognitive bias present in everyone, which can distort our perception of reality. This is a significant issue that deserves further examination. I address this in the online appendix.

Insightful Challenges

Impactful insights often encourage you to reconsider your thoughts and actions. When you change your perspective or behaviors, you're likely to view your product line system in a new light.

It is also important to consider how quickly you can generate meaningful and impactful insights. The faster you gain such insights, the sooner you can enhance the velocity of your outcomes.

An Insight-Generating & Sensemaking Machine

Product managers must provide guidance and set boundaries to help their teams form insights. These "enabling constraints" create a focus that aids in exploring the product line's architecture and the interplay of the system's parts.

A guided focus, rather than a laser-sharp one, fosters insight formation and facilitates the inclusion of those insights in adaptable strategic move charters. A flexible approach and versatile charters increase the chances that insights will be meaningful.

The approach must also help teams form valuable insights. This requires governance to develop and implement approaches that eliminate obstacles and barriers hindering insight formation.

A governance body must develop the capacity and skills to generate insights. This role is much more than a naïve directive to "go forward and create insights."

From Insights to Strategic Moves

Reasoning enables you to gain insights and connect them. However, we need strategic thinking to create an impactful series of strategic moves.[73] This approach assists teams in generating new ideas, considering the future, and planning their actions accordingly. Strategic thinking extends beyond simply generating insights and making sense of situations; it also supports the processes of creation and planning.

Generating Insights and The Accelerating Change Race

To address the simultaneous challenges of efficiency and change, teams must generate and utilize insights about their product line's system parts and architecture. Insights drive strategic moves that enhance a line's efficiency and change, while improving its Outcome Velocity.

This work begins by understanding the current and potential influences on a product line's parts and agents. It involves environmental scanning, research, knowledge probing, and "safe to fail" experiments. The work must emphasize the context of a product line and the associated knowledge domains.

Chapter 17 Summary

Generating Meaningful Insights

- **Insights are Crucial:** Many managers think insight generation is a creative task limited to a few people. However, it's a crucial aspect of managing product line systems.

- **The Fuel and Fodder:** Insights act as nutrients for product line systems, directing decisions and giving purpose to workflows. They are basic ingredients to form impactful strategic moves.

- **Personal and Organizational:** Big companies face unique challenges that demand an organizational approach to generating insights and converting them into a stream of strategic moves.

- **Two Focal Points:** Teams should adopt two focal points for generating meaningful insights: 1) the interplay of system parts and 2) the product line architecture.

- **Sensemaking and Reasoning:** Teams can make sense of insights by using different reasoning that ties the insight to the system and its performance.

- **Understand the Product Line:** Teams generate better insights when they understand the system well and aim for relevance over volume.

- **Continuous, Curious Learning:** Product line teams must adapt to changing situations and improve their system's efficiency. They must also remain curious about all aspects of their product line and its agents.

- **Safe-to-Fail Experiments:** Conducting experiments is an effective way to generate and validate insights. However, teams must design these experiments to avoid damaging the business or project.

Chapter 18

Insight & Sensemaking Affordance

Exploiting the Innovation Charter

This chapter examines how to generate insights that advance a product line. The next chapter discusses the leadership qualities and personal initiative required to execute the work.

O ld-school product managers and product line system managers share a common goal: both strive to do their best for their companies. However, the concerns they prioritize differ. Understanding these differences is crucial in participating in the Accelerating Change Race.

It's common for managers trapped in the single-product paradigm to focus on identifying and resolving the root causes of obstacles that might hinder cash flow or project speed related to a product. However, they often overlook insights into a product line system's behavior, components, agents, or structure. These topics remain outside their consideration.

Product line systems management differs from a one-product-at-a-time approach. A product line depends on insights and sensemaking about its current and future components, forces, enablers, and stakeholders. Without this effort, the line will struggle and ultimately fail.

Inviting Insights

Some people studying innovation believe that a unique culture is essential for gaining such insights. Generally, that's true. However, managers of complex systems must understand the environment and evaluate whether it facilitates or hinders the emergence of outcomes. This environment and its attributes are crucial. When the outward presentation of the trappings encourages you to engage in specific work and subsequently helps you complete that work, we refer to these attributes as the system's "affordance."

A psychologist defines "affordance" as what an environment offers or provides to an individual. For example, the affordance related to wealth creation and education differs between wealthy and impoverished communities. In design, an affordance is a visual hint that suggests what an object does and how it works. For instance, regardless of its shape, a door handle indicates that the door can be opened by pushing or turning it. This inherent affordance encourages you to use the handle to open the door. If the designer fails to provide the affordance or completely omits it, people may find themselves standing in front of the door, unsure of what to do.

In complex systems, the objects that influence the environment, the system itself, the actors, and the tools collaboratively create affordances for various actions. Our aim is to develop insight-generating and sensemaking affordances. We want to "invite" people to generate insights and make sense of them within a product line system context, specifically focused on improving their line's Outcome Velocity.

Affordance acts as an enabler that promotes insight generation and sensemaking, which subsequently leads to the development of strategic moves. We do not acknowledge such affordance in traditional single-product innovation and management.

Affordance, Behavior, and Culture

Please don't confuse affordance with culture. Affordance shapes behaviors that contribute to culture but is not culture itself. Our challenge is to ensure that the affordance we create doesn't have to contend with cultural forces pushing in the opposite direction.

Consider that a company's culture can foster management behaviors that resist change. A strict focus on time-to-market KPIs could be one of these behaviors. You may also find that an individual's performance review regards the time spent on generating insights as unproductive. Furthermore, strict budgeting practices resulting from an efficiency-driven culture might allocate no resources for this essential product line work. Organizations need to eliminate or counter these negative forces to promote insight generation and sensemaking opportunities.

Negative forces become apparent when you assess potential contributors. Inquire about what might prevent them from participating in generating insights. Compile their responses and categorize them by themes or common influencing factors. Once the factors are clear, the challenge lies in determining the best way to mitigate them. The solution will depend on the organization and the product line. While some may see this as a cultural shift, I perceive it as addressing specific forces and encouraging different behaviors to facilitate change-related efforts.

Insight-generating and Sensemaking Affordance

Established, mature companies often struggle to maintain a consistent flow of strategic actions that result in significant improvements in product line Outcome Velocity. Unfortunately, their typical approach is to delegate this responsibility to the product line manager or, even worse, to individual product managers. Many top executives somehow expect these talented individuals to solve everything on their own. However, that's an almost impossible task.

Instead, the solution lies in establishing practices that encourage the development and application of insights. Without intentional efforts to promote the formation of these insights, the emergence of outcomes will be slow and may even come to a standstill. In that case, teams rely on luck rather than intelligence to drive their product line.

Generating insights and interpreting them must enable teams to identify potential influences and provide the tools needed to analyze their effects on the system's components and the emergence of outcomes. The challenge lies in incorporating wisdom and foresight into these analyses.

Sensible insights are the genesis of strategic moves.

Affordance is not an invitation to innovate anything you desire, nor should it compel contributors to create complete product concepts. Instead, product line affordance should actively promote insight generation along with product line sensemaking. From this work, teams can then pursue the emergence of strategic moves.

The purpose of generating insights and making sense of affordances is to enhance a team's knowledge and refine the targeting and development of the line's strategic moves. Organizations achieve this targeting by embedding insights within potential charters[74] for products, platform-levers, and other system agents. It then becomes the responsibility of other system enablers to assist teams in vetting the charters and ensuring they are ready for investment and the roadmap.

PLSM Roadmap

I understand that many managers believe the sensibility and potential impact of certain insights become clear only after they are integrated into a strategic move. However, in PLSM, we model strategic moves before allocating resources and developing them.

The modeling requires teams to adopt innovation charters and use them to reflect insights and other knowledge. As explained in Chapters 17 and Appendices 5 and 7, the charter is a distinctive conceptual object that encapsulates specific insights and existing knowledge. Each charter serves as a target for a potential strategic move. It is neither the move itself nor a description of what the move may entail.

A charter is always connected to other parts of the system. Sensemaking then evaluates whether the move aligns with the line's roadmap and the cumulative effect of all strategic moves (outcomes).

Front-end Misguidance

Most traditional companies separate front-end work from development work. Front-end work focuses on delivering product concepts and does not emphasize non-product strategic initiatives.

Those involved in product development projects rooted in traditional front-end practices tend to focus intently on time-to-market. They often view newly formed insights as external factors that can delay delivery timelines. Consequently, the project management immune system may engage to exclude such insights from their work by either ignoring them or categorizing them in a "backlog." Traditional front-end work can hinder a product line's Outcome Velocity gains without managers being aware of it.

A major problem with the single-product approach is that it diverts resources away from more impactful strategic moves, shifting attention to daily product management tasks rather than generating insights. In this old-school approach, the only way to translate significant insights into actions is not through product management; it's through top management. However, most practical thinking occurs at the lower and middle levels of the organization.

PLSM's primary role is to help teams tackle these challenges. It removes delays and bottlenecks that obstruct the formation and implementation of valuable insights. This role is distinct from using Agile, Lean, or the Theory of Constraints to address project bottlenecks and constraints.

From Data to Insights to Moves

In product line systems management, each product line team's most crucial task is to generate impactful insights that exceed those of competitors. Subsequently, team members and contributors must leverage all available knowledge, including these insights, to devise potential strategic moves based on them. They must then refine these potential moves to enhance the overall strategy.

Although I've described the necessary work in a linear progression, the reality is more of an ongoing iterative cycle from which moves develop. The value-compounding nature of the strategic move stream is a crucial aspect of product line systems management. Similar to insight-generation affordance, it distinguishes product line systems management from traditional single-product management. See Figure 18.1.

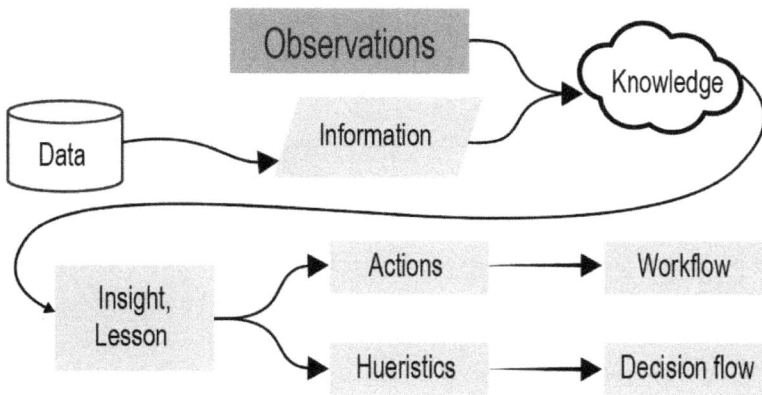

Figure 18.1 Knowledge and information flow leading to Strategic moves (work and decisions).

Insights are essential, but they only become valuable when they are integrated into or influence a strategic decision. And it's valuable insights that are necessary to transfer notable efficiency and change into a product line.

Creating and Nurturing Affordance

Affordance offers encouragement to engage in and perform activities that generate insights and facilitate sensemaking. It prompts managers to undertake significant work. As you create such an affordance, consider the potential elements outlined in Table 18.1 below.

Table 18.1 Possible Affordance Elements

Affordance Element	Discussion
Establishing an insight-generating and sensemaking portfolio review process	Encouraging, resourcing, and communicating partial and fully thought-out investigative ideas
Helping actors and potential actors to learn PLSM context, concepts, components, and principles	Understanding PLSM and embracing a product line's architecture, parts, agents, actors, forces, and constraints is necessary to enable insight-generating and sensemaking affordance. You will fall short in these areas if you do not understand the system.
Building an understanding of Value Creation and Outcome Velocity improvements	Requisite knowledge
Teaching the tools of Insight Generating and Sensemaking	Requisite knowledge
Teaching and exploring your product line, system stack, and potential variants	Requisite knowledge
Sharing an understanding of the product line key part, the Roadmap, and the system assemblage map	Foundational knowledge to explore potential moves
Learning about the System's Architecture and Archetype, Pivots and Transformations, and the Interplay with business models and product line strategies	Foundational knowledge to explore potential moves
Exploring how to create, improve, and use Innovation Charters for products, platform-levers, and system agents.	Requisite knowledge

Affordance Element	Discussion
Teaching and exploring the role of System Enablers and Constructors	Requisite knowledge
Teaching and exploring System Flows, Flow Constraint Mitigation, and Flow Resonance	Requisite knowledge
Setting up and carrying out Charter reviews, Onboarding Charters, Roadmapping, and Charter Resourcing	Requisite knowledge
Guiding actors in understanding Strategy as an Enabling Constraint, setting up practices to change the product line strategy.	Foundational knowledge to explore potential moves
Building Clarity on Current Product Line Strategy, Communicating all aspects of the product line system.	Requisite knowledge
Creating informal Support Groups on technical, marketing, or organizational topics related to product system agents and their interplay	Helpful supports
Setting up performance appraisals that capture insight generation and sensemaking, agent advancements, and forming and carrying out strategic moves	All contributors and actors are expected to get involved and produce.
Training and reinforcing Distributed Leadership and Individual Initiative	Requisite skill development

Affordance Element	Discussion
Fostering transparency of work and recognition for work, and contributions.	Visible recognition
Setting monetary and non-monetary rewards for thinking and work.	Incentives that motivate behaviors
Fostering top leadership participation and involvement / setting expectations, and encouragement	We expect involvement/participation, not specific outcomes
Resources Tools, Support, Templates: Chater Templates, Sensemaking (Sensemaker™- The Cynefin Company)	Interface of Sensemaking with PLSM processes, practices, and Decision Flow
Providing Money when needed	Set up Insight-generating and Sensemaking as Portfolio
Ensuring actors have the Available Time and Bandwidth to carry out individual initiatives.	Consider 3M's allowance of 15% of an individual contributor's time.
Making Technology and Marketing Expertise available to individual contributors	It's hard, if not impossible, to work without resources
Creating Shared Learning Forums on specific topics;	Moving deep learning into the organization
Continually recasting and strengthening all PLSM affordances	PLSM insight generating and sensemaking affordance must not be static

First Remove Blocks

A key principle in organizational change is to remove obstacles and barriers that hinder desired behaviors as much as possible before introducing factors that encourage those behaviors. For example, before launching a bonus plan for submitting ideas, first understand why people might hesitate to do so without a bonus. Then, find ways to remove or lessen this obstacle. The reason is that when drivers encounter obstacles and barriers, the resulting turbulence can disrupt or derail the initiative.

Affordance encourages specific behaviors and acts as a driving force. Therefore, it's wise to identify and eliminate any obstacles or forces that reduce the effectiveness of the affordance. The remaining elements of single-product thinking, along with the supporting processes and practices, are often the main culprits. Examples include time-to-market KPIs, outdated performance reviews, and excessively slow gated decision-making processes.

I advise any organization or team aiming to establish insight-generating and sensemaking capabilities to first identify their obstacles and develop a plan to eliminate or reduce them. The best people to do this are those who you wish to carry out the insight generation and sensemaking. From there, top managers must get involved in helping to remove the obstacles and begin implementing elements of the desired affordance. The online appendix provides a more detailed look at setting up insight generation and sensemaking affordance.

Responsive Sensibility

Responsiveness is the capacity to act swiftly when an insight suddenly becomes evident. These actions and decisions generate or maintain value. The goal of responsiveness is to exploit insights rather than to create new ones.

The good news is that merging insights can lead to new insights, which may result in value-adding actions and decisions.

The most significant insights emerge at the intersection of technology and markets. Furthermore, it is often observed that the technologies companies utilize to make major strategic moves are available to most competitors in an industry. What truly matters is the innovative application of existing technologies that creates value for customers.

Grand insights aren't about groundbreaking technology or a newly discovered customer need. Instead, the spark of genius lies in how a product line team can effectively utilize the technology and align it with customer needs. Value only emerges when a team can meaningfully match technologies with customer needs. This is both a process and a flow challenge.

Brilliance emerges from effectively manipulating and utilizing the product line system. Teams and their organizations must do everything possible to cultivate this brilliance. It is the fuel that powers the line's Outcome Velocity and capacity for change.

Appendix Case Study 2 presents an interesting case study of DANSLENS and Johnson & Johnson. It explains how Johnson & Johnson derived value from their use of the technology, rather than from the technology itself. The value they gained greatly exceeded what others believed the technology was worth.

When new technology emerges, teams must understand its impact on their product line system. The resulting insights offer crucial guidance for product line strategies, influencing changes like extending, recasting, pivoting, or transforming the product line.

Insight Affordance and The Accelerating Change Race

Insight affordance encourages and assists people in generating insights about a product line system. It provides managers with the tools and capacity to integrate their knowledge and sensemaking abilities.

Affordance paves the way and eliminates obstacles for managers to generate insights that drive strategic actions. These actions are essential for running the Accelerating Change Race.

Chapter 18 Summary

Insight, Affordance, and Strategic Moves

- **Systemic Approach:** Product line systems management diverges from traditional single-product methodologies by focusing on the system's interconnected components, forces, and actors.

- **Insight-Generating Affordance:** In product line management, "affordance" refers to creating an environment and providing the tools and capacity that invite and help contributors to generate insights that contribute to meaningful strategic moves.

- **Cultural Impediments:** Organizations must identify and mitigate cultural barriers that inhibit insight generation, such as rigid adherence to time-to-market metrics or efficiency-driven resource allocation.

- **Strategic Move Genesis:** Insights are the foundation for strategic moves, essential for enhancing product line performance and adapting to market dynamics.

- **Innovation Charter Use:** Employing innovation charters as conceptual frameworks helps encapsulate insights and delineate potential strategic moves within the context of the broader product line system.

- **Traditional Development Pitfalls:** Conventional product development practices often marginalize new insights, potentially stifling innovation and impeding product line Outcome Velocity.

- **Technology Application:** The most consequential insights frequently arise from novel applications of existing technologies rather than the advent of new technologies.

- **Continuous Insight Generation:** Generating insights and translating them into strategic moves is iterative and ongoing. The work helps form a value-compounding stream of strategic moves.

- **Responsive Adaptability:** Effective product line management requires the capacity to swiftly act upon emergent insights and the skills and resources needed to explore and implement these revelations.

- **Efficiency-Change Breakthrough:** Insight affordance plays a crucial role in addressing accelerating change by providing managers with the tools and capacity to generate insights that aid operational and product line change.

Chapter 19

Distributed Initiative & Leadership

How Agency Exploits Uncertainty

Individual contributors need to choose which components and agents of the system to explore for insight and sensemaking. Moreover, those involved in this work should seek help from others. Managers who start these efforts should be intelligent, resourceful, and, most importantly, self-motivated.

Leadership vs. Initiative

Human Resource professionals quickly emphasize that leadership is the most important factor influencing all aspects of business, especially in large organizations. When situations seem complex, the solution often depends on leadership. However, the preferred approach is to delegate leadership and decision-making to the lowest possible level. The HR community calls this distributed leadership.[75]

Individual contributors play a crucial role in advancing complex product line systems. Their challenge is to generate insights, interpret these insights within the product line context, and apply the newfound knowledge effectively. Usually, this is not a solo effort. Ad hoc teamwork is required. Such collaboration demands distributed leadership. This role is carried out by individual contributors who, without formal authority, take the initiative to mobilize resources and drive specific sight-generating and sense-making initiatives.

Product line systems management involves making choices and judgments, part of decision flows, at the "lowest level of granularity" in the system. It also requires investment control at the overall systems guidance level. However, what's most important is not leadership at the lowest level. This practice needs initiative, resource support, and effective communication. Before we move to distributed leadership, let's look at these three contributions. The most important is initiative.

Why Initiative?

We are never completely sure about a product line's results or what affects their development. This is typical of a complex system. Such uncertainty forces product line teams to take a unique approach to improve their practices. It's also a big turnoff for some managers. For these people, the affordance we've discussed may not appear encouraging.

The affordance stems from the availability of resources, structural support, and purposeful encouragement. But it works only through self-directed efforts. This approach contrasts with top-down command-control in which that most old-school managers grew up. Instead, it provides the permission to "go do as you see best." The path is specific to the product line system and its current and potential components. You'll notice that this approach leverages self-directed organizations. But the initiative of an individual, preferably many individuals, is a prerequisite to its effectiveness.

Someone needs to step up and lead possible explorations. Gaining meaningful insights and generating potential strategic moves won't happen without initiative and guidance. A person must spearhead the effort. Since this initiative operates outside a command-control hierarchy, it requires independent leadership. This independence is known as Distributed Leadership. It spans various roles and functions, regardless of pay levels.

Self-Starters

Independent self-starter thinking has always been encouraged in innovation. Many companies copied 3M's "15% Culture," which fosters contributors to dedicate 15% of their time to developing new product ideas they believe are worthy. This "free time" approach gained popularity in the 1990s but stalled as lean methods pushed for greater efficiency.

PLSM affordance differs from the 15% Culture. While still pursuing 10 to 15% of people's time, PLSM affordance seeks to gain insights and develop strategic initiatives related to the product line and its components. For most companies, but not 3M, the 15% Culture encourages any new product opportunity that "fits" the company. It tends to be single-product oriented.

One important aspect of 3M's approach is often overlooked by companies attempting to replicate the 15% Culture: 3M emphasizes platforms over products. The company does not use the term platform-levers; rather, it seeks new platforms that support multiple products and encourages contributors to think of new products that utilize their existing platforms. This platform orientation serves as an enabling constraint that enhances emergence far more than the general idea of "fits the company." 3M's focus on platforms is evident in the title "platform manager" given to many managers. This approach prioritizes multiple products instead of just a single product.

Resource Support

Examining how the system's components interact is crucial for influencing a line's Outcome Velocity. The task involves identifying potential strategic moves that could enhance the line and uncover any limitations or obstacles to implementing these moves. The emphasis is on increasing customer satisfaction, competitiveness, and cash flow.

The most important aspect of product line system affordance is the funding it provides. Individual contributions can only go so far before additional resources become essential. This funding is separate from what is designated for development projects and strategic initiatives. An affordance steering team is tasked with allocating the funds.

The total affordance budget should be between 0.5% and 2% of annual revenue, depending on the growth rate of the markets served. Additionally, I recommend not starting such efforts with less than 0.25% of revenue funding. This is for insight-generating and sensemaking work, not the projects and strategic moves that emerge. I recognize that this is a significant amount of money in some companies and may seem to lower cash flow contributions to the line's Outcome Velocity. Nonetheless, it's table stakes for staying in the Accelerating Change Race.

The outcome of these efforts is a collection of new and valuable insights alongside a portfolio of potential strategic moves. Regardless of what arises, every team leader and participant will receive 360-degree feedback on their contributions. The teams will then incorporate this feedback into each contributor's annual performance review.

A very important aspect of affordance is gathering and sharing insights from everyone's efforts.

Communicating and Integrating Strategic Moves

An Example

Allow me to present a fictional example that demonstrates Distributed Leadership in PLSM.

Jennifer is a marketing manager at a large company and is working on a major product line. Through her daily work and discussions with others, Jennifer identified issues and opportunities related to a specific market segment. She took it upon herself to conduct some secondary research and found nothing that convinced her to halt further efforts.

The self-directed initiative was formed when Jennifer recognized opportunities for one of the company's technologies to assist in the market segment. However, she also identified customer issues that might be more significant than she realized. Jennifer understood that she was dealing with various parts of the product line system. As she explained, these included a noun-based segment, two technologies, and the potential of a supplier's technology.

Jennifer's primary responsibility was to recruit three additional members for her team. Two of them were knowledgeable about the technologies, while the other was a supply chain expert. At the request of the supply chain expert, Jennifer agreed to broaden the scope of their exploration. This secured the participation of each expert.

The team of four, led by Jennifer, defined their mission regarding both current and potential components. They aimed to explore the possible impact of the system components on Outcome Velocity in the mid- and long-term. Their goal is to develop insights and formulate potential strategic moves.

The team's first task was to work independently to compile knowledge, insights, and lessons they believed would be beneficial. They also developed a list of assumptions that the company appeared to embrace regarding the parts. Jennifer then created a new Slack channel for the project and invited the others to join. She estimated that the project would take about 90 days. There was no formal project plan.

After a few weeks, the team shared their insights and identified areas for further exploration. They believed they needed deeper understanding of customer needs and behaviors, along with advancements in supply chain technology. They interviewed an external expert who specialized in a specific supply chain technology, along with several customers in the targeted segment.

In the following weeks, the team exchanged numerous messages and documents. The technical members conducted only one experiment addressing capacity issues related to technologies that served as essential building blocks in the product line system. Plus, Jennifer developed an interview guide and organized 20 potential customer interviews.

The team, led by Jennifer, subsequently delivered a presentation to the Product Line Affordance Steering Team.

Jennifer and her colleagues received the funds, but the steering team also recommended that they bring on a new member with business development and sales skills.

Jennifer's story suggests she has considerable support for her work. One would hope this support is directed towards Jennifer, allowing everyone to enjoy the same level of backing. Such support is a cornerstone of insight generation. The broad support at the lowest level of granularity indicates a robust capacity for generating insights, but only if it is paired with resources and a mechanism for communicating challenges and successes at this lower system level.

Not all initiatives should receive full support and resources; that would be nonsensical. We need a weed-and-feed approach that regulates these initiatives, along with a communication mechanism to share insights and lessons learned. The main point is for individuals to take initiative, not to assume a leadership role.

Distributed Leadership, Initiative, and The Accelerating Change Race

Generating insights and making strategic decisions are challenging tasks. Managers must have the skills to guide colleagues in merging ideas and efforts. This starts with motivation and personal initiative.

The job isn't to finish all the work needed to move a product line forward. Instead, it's to develop and refine insights and to shape and reshape ideas. It's about turning these ideas into recognized projects that intentionally lead to strategic moves.

Initiative and distributed leadership are crucial for helping a product line team keep up in the Accelerating Change Race. They promote freedom of thought in the "knowledge trenches" and guide efforts toward meaningful results.

Chapter 19 Summary

Distributed Initiative

- **Distributed Initiative Need:** Uncertainty in product line outcomes demands a unique approach. Distributed Leadership, spread across roles and functions, is essential for independent initiative and guidance in exploration.

- **Affordance Steering and Funding:** Affordance involves steering self-directed efforts to uncover strategic moves and limits in a product line. Funding for affordance, allocated by a steering team, is crucial for driving a product line's Outcome Velocity, typically ranging from 0.5% to 2% of annual revenue.

- **Example of Distributed Leadership:** Jennifer, a marketing manager, begins a self-directed exploration of several product line parts. She recruits colleagues with diverse expertise, leads the exploration, and secures funding for experiments and research, showcasing how Distributed Leadership works in Product Line Strategy and Management (PLSM).

Chapter 20

Organizational Structure & System Alacrity

Combining Hierarchical and Lattice Structures
Adding Insight & Sensemaking Affordance

E verything changes, and it's happening at an increasingly rapid pace. That's the essence of the Accelerating Change Race. And when technologies and markets change, the product line and all its supporting elements must change as well.

Product line teams must embrace change and adopt new practices, processes, and structures. Such changes help to reform a system's behavior. This chapter discusses structural changes that enhance a product line's performance as they join the Accelerating Change Race.

Archetype, Strategy, and Structure

All product line teams face significant challenges in dynamic and competitive business environments. First, they must determine and establish a product line system archetype. Then, they must develop a strategy and structure. As the business maxim goes, "Structure follows Strategy."

An organization's structure is a key enabler of a product line's performance. For it to be effective, the structure must align with the product line system, including its archetype, strategy, roadmap, and system assemblage map. It also needs to support the system's flows and enablers. Refer to the Enablement Pyramid in Figure 20.1.

When a product line strategy shifts its focus from solely efficiency to a combination of efficiency and change, the components in each layer of the pyramid must be adjusted.

I am not suggesting that an entire business needs to change. Instead, changes to product lines require teams to assess and adapt the system's behavior. For instance, when platform-lever types change, the system strategy must also evolve. This suggests that the structure should shift to prioritize the new platform-lever.

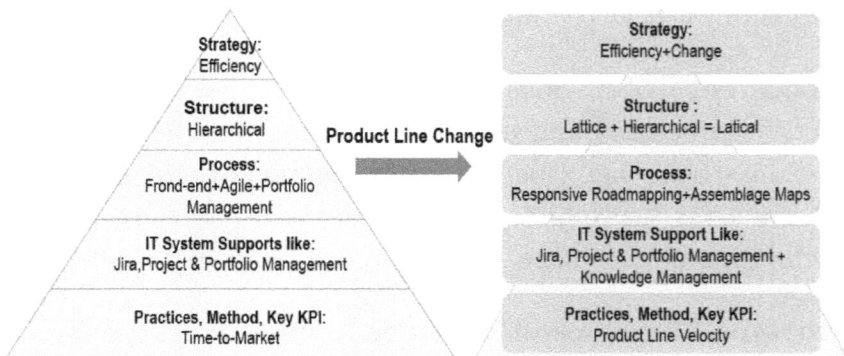

Figure 20.1 The Product Line Enablement Pyramid. When a product line changes its archetype or strategy, the organizational structure may also need to change.

272

Let's explore how companies can alter their structure to improve, rather than hinder, a product line's performance and adaptability.

Organizational Structure and Change

An essential principle of business management is to create a deliberate focus on what you want to improve. To enhance something, concentrate a strong focus on it. One way to do this is by establishing an organizational structure with dedicated roles, each with specific responsibilities and authority related to the improvement goal. This focus-driven approach supports most hierarchical (ladder-like) structures. The challenge for PLSM is that improvement always involves efficiency and change.

The complexity of a product line involves many interconnected parts, forces, and constraints that influence its behavior and performance. Many factors may need close attention. However, the hierarchical structure cannot focus on everything. Instead, a narrow focus encourages higher levels of command and control rather than flexibility and responsiveness. The efficiency-change tug-of-war presents a major organizational challenge for product line systems.

Alacrity in the Trenches

The military teaches an important lesson about managing combat. They establish their presence on the battlefield to let corporals and sergeants lead ground maneuvers and actions. This method consistently produces better results against enemies stuck in a top-down hierarchy.

Non-commissioned officers recognize the importance of flexibility and responsiveness in combat. They neither need nor want a commander dictating every detailed action. This reflects how Western-trained Ukrainian forces counter the hierarchically controlled Russian army.

However, skilled battlefield actions always involve a higher-level strategy. To achieve this, commanders must communicate information and insights quickly. But that's not what happens in the military's daily non-combat operations. Standardized routines take over for "normal" operations. These activities aim to improve operational efficiency.

Organizational Contingency Theory

Researchers have conducted extensive studies on a company's structure and its relationship to performance. Initiated after World War II, this research aimed to identify the optimal structures for enhancing productivity. The effort resulted in the development and validation of "Organizational Contingency Theory."[76] It initially focused on top leadership but shifted to organizations as the concept of distributed leadership gained traction.

This empirically supported theory posits that the optimal management approach is contingent upon the situation.

The best course of action depends on both internal and external factors. A top-down approach is most effective when these factors are stable. In contrast, a bottom-up approach that encourages decision-making at the lowest level is more effective when circumstances change rapidly.

The issue has been that top-down hierarchical structures conflict with a bottom-up approach. The solution has been to encourage senior leaders to adapt their style to support and motivate their subordinates in implementing distributed, bottom-up initiatives and decision-making. Success, we're told, depends on top management leading the way. Unfortunately, relying solely on leadership to "foster" the bottom-up approach has not proven effective.

Most companies still follow a command-and-control hierarchy. They try to be flexible by using cross-functional teams and fast project management. However, this approach falls short. Encouraging leadership alone can't keep up with the need for the notable product changes today.

Lattice Organizational Structure

An organizational structure called the Lattice structure was a popular management topic in the 1970s and 1980s. In this setup, individuals manage and direct their own work without a hierarchy of bosses overseeing what gets done. This structure seems to work well for product lines. In complex product line systems, changes happen through the collective efforts of individual contributors. A Lattice structure supports this kind of distributed work.

Interest in lattice structures declined as the efficiency-focused movement gained momentum in the 1990s. Only one major company still operates within a lattice framework: W.L. Gore and Associates.

Understanding Gore's lattice structure and its functionality is useful. The challenge, however, is that Gore, being a private company, keeps its organization secretive (which makes sense). Babson Professor Jay Rao's case study provides an excellent summary of W.L. Gore's approach.[77] In discussing the case with Professor Rao, I learned that he never interacted with any Gore employees. He created the case entirely based on publicly available information. My experience with W.L. Gore and conversations with Gore employees about the case study have led me to support Professor Rao's work.

A few points can help you understand a modern Lattice Organization. A pure lattice organization has no hierarchy. Everyone works in the way they believe is best for the organization. However, the concept of "no hierarchy" does not apply to a modern Lattice Organization. Today, key managers hold hierarchical titles (e.g., Directors, VPs, etc.) that define their responsibilities for profit, loss, and performance. However, individual contributors report to two or three superiors from various functions, from whom they receive guidance and broad directives. Each associate collaborates across multiple teams and small ad hoc groups throughout the year. Most also participate in traditional (not distributed) product development projects, while some may engage in non-product-related initiatives.

An associate's annual performance review includes multiple 360-degree assessments of their managers, teams, and temporary groups they work with. These reviews gather feedback on job duties and team input. I imagine that nearly every Gore Associate got report cards in elementary school that said, "plays well with others."

Lattice Flexibility

W.L. Gore encourages all associates to form and manage ad hoc teams as they see fit and to participate in project teams created by others. Performance appraisals reflect contributions to each team. These evaluations consider each person's tangible and intellectual input, as well as how much their efforts have advanced the product line. For Gore, it's clear that this approach fosters contributions to products and innovation that might not otherwise be possible.

W.L. Gore has spent over 50 years fostering an environment that encourages distributed leadership and independent initiatives. From my experience working with their Medical Device business unit, I observe that they do not operate strictly as a lattice organization. They can't. If they did, profitability would be impossible. However, they cultivate a culture and numerous practices rooted in innovating like a lattice organization.

Over the past few decades, many people have considered the Lattice organization impractical for any business seeking growth or scaling. There is some truth to this belief. However, no company, not even W.L. Gore, uses a purely Lattice structure. Instead, Gore's structure is a hybrid of Lattice and Hierarchical, which I call "Lattical." It's a mash-up of the two main structures. You will see that it's feasible and has proven successful. See Figure 20.2.

W.L. Gore demonstrates that the two structures can coexist. Efficiency and change can thrive within a single framework. However, the glue that binds them together is the crucial insight-generating and sensemaking affordance discussed in Chapter 18. Encouragement from top leadership serves as another key aspect of this affordance.

Please do not assume that, based on what I am discussing, W.L. Gore endorses or carries out everything I've shared in this book. I am merely discussing the company's organizational structure and the affordance they created for insightful innovation.

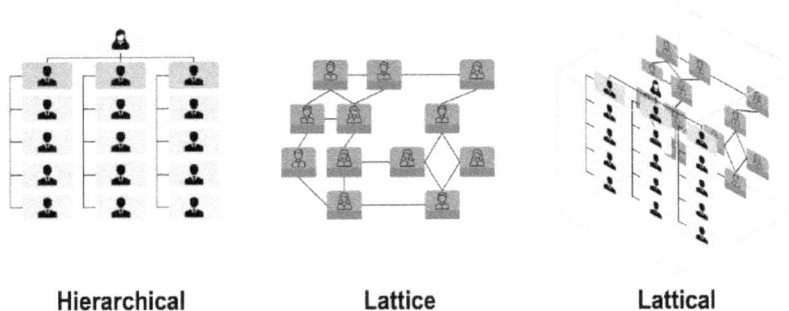

| Hierarchical | Lattice | Lattical |

Figure 20.2 *A Lattical Organizational Structure combines Hierarchical and Lattice structures. The combination plays out in reporting relationships, roles and responsibilities, and performance appraisals.*

In the new Accelerating Change Race, product lines require a flexible bimodal structure. A strictly hierarchical structure boosts efficiency but fails to support change. In contrast, a purely lattice structure encourages change but sacrifices efficiency. The ideal structure should combine both models to address efficiency and change simultaneously effectively.

Distributed leadership roles in the Lattical structure do not carry the responsibility of delivering key system components such as new offerings, platform-levers, or enabling agents. That responsibility lies with the functional and cross-functional teams typically found in a hierarchical setup. In PLSM, independent initiatives and distributed leadership are tasked with embedding knowledge and insights into charters for products, platform-levers, and agents. (See Chapter X and Appendix X for more on charters.) Teams then assess potential charters based on their ability to advance the product line before adding them to their project portfolio.

The methods and practices of transferring meaningful insights— those derived from independent actions and distributed decision-making—into standard project and portfolio

278

management are vital for the support that sustains the Lattice structure. Without a destination for insights and a way to communicate them, distributed initiatives and leadership lack direction.

An Intense Lattical Focus

Earlier, I explained how a product line's strategy acts as an enabling constraint. I also emphasized that teams must focus intensely on the most critical 10% of the system's parts and agents. Platform-levers, which are essential to the system's archetype, are always part of these vital components and demand this focus.

Teams achieve sharp focus by assigning roles and responsibilities to skilled managers who oversee the system's key components and agents. These roles are integrated into the overall structure, and teams must establish cross-functional processes to refine the system's flows and flow resonance. This structure-enhanced focus boosts system performance and responsiveness. Individuals (system actors) then take responsibility for ensuring efficiency and driving change.

Big Gains: Bridging / Active Affordance

As forces and constraints change, temporary structures or flow enablers might be needed. I call these bridging enablers. Think about how organizations often run single events to support the system, such as constraint mapping or setting project priorities. Bridging is a way to get tasks done while learning as you go. A bridging enabler is crucial because it increases the flexibility of structures and processes.

Bridging can promote insight generation and sense-making from both large system-wide functions and individual actions. When actor-managers engage in bridging activities within product line systems management, they play a crucial role in understanding the purpose and value of the bridging enabler. This involvement helps establish more enduring enablers that drive insights and sense-making within the system's flows.

As discussed in Chapter 18, insight-generating and sensemaking affordances enable distributed leadership to take hold in a lattice structure. It is surprising to see many leadership experts claim that top corporate leaders must promote in-the-trenches distributed leadership without providing the tools, templates, or structures needed to do so. For example, I cannot facilitate distributed leadership without training in insight generation, sensemaking, and product line systems management. It also requires companies to implement team-oriented 360-degree performance reviews, like those that W.L. Gore conducts.

Centers of Excellence

Product, technology, and market knowledge come from accumulated learning. Knowledge turns into understanding through experience and judgment. In product line management, gaining insights involves figuring out how to use this understanding to improve a product line's Outcome Velocity. Ongoing improvements in Outcome Velocity begin with deep knowledge. Therefore, it is crucial for a company's structure to intentionally focus on developing specific knowledge.

In Figure 9.2 in Chapter 9, you will see that I present a product line stack that includes "competencies" under technology building blocks. These are organizationally designated Centers of Excellence or functional groups that gather knowledge and information to develop a deep understanding of a critical technology, marketing, or PLSM topic essential to the product line. Their role is to explore and experiment with all possibilities related to the general technology and adjacent technologies concerning one or more product lines.

You will find a Center of Excellence dedicated to core technologies related to building blocks and platform-levers, as well as advanced topics like AI, automation, and blockchain. Marketing topics may include Jobs-to-be-Done, user experience, and customer retention. PLSM topics might cover archetype transformation, constructors, and assemblage mapping.

Centers can include groups within R&D or marketing functions. They can also be cross-functional teams that operate across the organizational hierarchy. A Center of Excellence is directly related to the development of the product line system. Each center's specific focus, assigned personnel, and the management group they report to are decisions made by business leadership aligned with a strategic vision.

Organizational Structure and The Accelerating Change Race

An organization's structure can either impede or improve a product line's performance and flexibility. A hierarchical setup typically resists change and emphasizes efficiency, whereas a lattice organization encourages individual initiatives and decision-making at lower levels. The lattice structure supports change rather than focusing solely on efficiency.

We know, however, that the two structures can unify into one, improving efficiency and supporting change. This merged structure highlights insight creation and sense-making while keeping a focus on standard operations.

To make the unique structure effective, organizations require proper insight-generation and sensemaking capabilities, along with a supportive practice that helps incorporate insights into PLSM charters.

Chapter 20 Summary

Organizational Structure and System Alacrity

- **Old-School Approach and Challenges:** Experts suggest standardizing practices and creating organizational structures, but the fast pace of change makes these approaches ineffective. Product line management must change its practices and embrace fresh approaches and structures.

- **Structure and Strategy Enablers:** Management teams must align strategy and organizational structure. They do this by recognizing that "Structure follows Strategy" and shifting from an "Efficiency-only" to an "Efficiency-plus Change" approach.

- **Organizational Structure and Product Line Changes:** Structure enables a higher-intensity focus on critical system parts and agents. The specific parts and agents on which to focus depend on the strategy. Leadership decides which elements need focus by setting the line's strategy.

- **Lattice Organizational Structure:** The lattice structure was popular in the 1970s and 1980s. It emphasizes individual direction and work management. Today, W.L. Gore is the only large company with a lattice structure. It encourages flexibility, distributed leadership, and ad hoc teaming.

- **Lattice Flexibility and Hybrid Structures:** W.L. Gore's lattice structure is not purely hierarchical. It also incorporates hierarchical roles and responsibilities. A Lattical structure combines lattice and hierarchical elements. It has proven practical and effective. It works because Gore also has insight-generating and sensemaking affordances in place. Such an affordance includes multiple 360 performance reviews of individual contributors. Across all W.L. Gore business units, the Lattice structure and affordance seem more geared toward single-product management, not multi-product management.

- **Lattical Structure for Product Lines:** Product lines benefit from a flexible Lattice structure. It focuses on the line's most critical components. It also enables insight-generating specific to a product line's key parts and agents. This work is necessary to drive efficiency and change.

Chapter 21

Enablers & Constructors

Boosting Flows and Creating Key Parts

E very system has "enablers." They're like backstage crew members—essential to the show, but not the principal actors. These system agents assist product line teams in addressing the two major challenges of the Accelerating Change Race: constantly improving a line's Outcome Velocity and building and maintaining a line's Change Capacity. It's unlikely that a team could tackle either challenge without enablers.

System enablers are practices, processes, environments, tools, and assets that serve two key roles: enhancing a system's flows and facilitating the creation and modification of system components. First, let's explore the enablers that can improve a product line's flow.

Boosting Flows

Consider the four flows in product lines discussed in Chapter 7: workflow, decision flow, information flow, and outcome flow. Any practice, method, or investment that enhances one or more of these flows is considered a system enabler. For instance, agile methods and staged development processes support workflows, while portfolio management aids decision-making. Table 21.1 illustrates several potential flow enablers along with their associated software applications.

Table 21.1 Common Flow Enablers

Potential Enablers	
Assemblage Deconstructing and Mapping Tools (stacks and change maps; Cynefin Co's Hexi Tool Set) Include Constructors.	Insight Generation and Sensemaking Affordance Manager; Signal / Weak Signal Detection; Sensemaking Tools (Consider Cynefin Co's SenseMaker tool)
Product Line Roadmapping Practices and Tools (Stack Maker)	Constraint Mapper / Type designator; Energy and Time mapper
Insight Generating, Probing, Combining, Building, and Tracking	Agent Change: Energy Time Impact analysis, System Coherence
Strategic Move Formulator/Planner; Energy vs Time; Furturecast ST and LT Velocity Impact; Challenges and Approaches	Ecosystem Coordinator / Communicator
Archetype | Business Model | Business Strategy Evaluator and Monitor, and Futurecasting)	PLSM Dashboard / Communicator - Velocity tracker
Emergence Facilitator	Continuous Velocity (Delphi / Foresight Aggregator)

Potential Enablers	
Constructor Creation Facilitator: Framing, Blueprinting, Planning, and Building	Decision / Judgment PLSM Principle Alignment Tracker
Constructor and Enabler Project & Portfolio Management	Training / Learning PLSM 101 through Advanced Team Workouts
Resonance Evaluator (Decision, Work, Information Flows)	Product Line Archetypes, Business Models, and Strategy | Deep Analysis and System Behavior Change/Impact Monitor
Alpha-Omega Product Portfolio Management (includes innovation charters -PICs, PIDs, PIMs, and retirements)	Innovation/Strategic Move/Charter Management: Pre thru Invested
Platform-Lever Monitor Health, Factors: Speed of Development and Bang for the Buck, LC Age	Agile, Stage-Gate Delivery Processes
Integrated Roadmapping, Portfolio Management, Sensemaking	Orderly System Root Cause Analysis /Roadblock Toolkit
Scenario Back-casting Tool & Process	Rapid orderly hindrance mitigation (TOC, Bottleneck/Constraint Analysis and response)
System Change Capacity Evaluator, Futurecasting, and Tracker	

Some enablers facilitate the interaction of a small set of system parts, constraints, and forces that influence only one flow. Other enablers may affect the interaction of many agents and influence multiple flows.

Enablers do not change the system's fundamental behavior patterns or archetypes. You can add or remove enablers, and while this may affect the system's performance, it will not alter the system's archetype. However, the reverse is not true. You must adjust the enablers in your system if you modify the system's archetype.

Multi-Product Enablers

Every product line system has constraints that affect its flows and limit its performance. When teams employ enabling tools or practices, such as lean methods or the Theory of Constraints, to tackle one constraint, other constraints tend to arise. The new "top constraint" suggests that the system has greater potential to offer. However, significant concerns arise that teams must consider. Does the cost of mitigating a constraint outweigh the short-term efficiency gains, or is there any improvement in the system's Change Capacity? Does addressing the next constraint complicate modifying the line's architecture? These topics capture the essence of efficiency inertia described in chapters 1, 5, and 18.

Most enablers in use today were not designed for complex multi-product systems. Instead, they focus on single-product orderly systems and improving efficiency. They do not help build Change Capacity within a product line.

This situation can take center stage when a team aims to enhance its product Outcome Velocity and adopts a multi-product systems management approach. Several of the enablers embedded in an organization's practices need to be reimagined and reconstructed.

Bridging Enablers

An enabler can also be temporary and improvised. Teams use temporary enablers to provide benefits and fill gaps. As they do, a more permanent enabler may be developed. I refer to these as "bridging" enablers.

An example of a bridging enabler is an impromptu meeting of the product line team to rank projects that support a specific technology. They might use a tool that integrates various spreadsheets or datasets through Microsoft's Power BI. If the tool proves effective and beneficial in other scenarios, it may be more prudent to incorporate its functionality into an enterprise system to facilitate its application and ongoing updates. Nevertheless, the initial use remains a bridging enabler.

Identifying and advancing enablers is the responsibility of the product line team, individual contributors, and a governance body. However, it is crucial to recognize that insufficient investment in enablers can impede a line's Outcome Velocity.

While enablers will naturally emerge as the system seeks to adapt and survive, proactively creating and integrating enablers is more effective in addressing a line's efficiency-change dilemma. This approach helps teams avoid delays in making critical strategic moves. To achieve this, however, the system needs an enabler of enablers. I refer to this grand enabler as System Assemblage Mapping. I discuss this important enabler further in the online appendix. There, you'll read about what it is, how it functions, and why I attribute much of the credit for this special enabler to Dave Snowden, the expert in complexity theory.

Constructors: Creating and Recreating System Parts and Flows

First, let's examine a special type of enabler that can significantly enhance a product line system. When an enabler directly creates, modifies, or automates a flow or builds a part without acting as an agent within the system, it is referred to as a "Constructor." It focuses on the assemblage of the system, not its daily operation. This unique type of enabler, now empowered by machine learning and artificial intelligence, has important implications for managing and evolving complex product line systems.

The landscape for investing in system enablers has undergone significant changes over the last few decades, although within the context of single products. Many companies have made remarkable advancements, especially in their methods, practices, processes, and support systems. We observe these advancements through Agile methods and Jira by Atlassian, as well as with portfolio management tools such as Sopheon, Planview, Planisware, and other company-specific toolsets. These enablers have facilitated and improved the flows that contribute to product innovation and management. Not surprisingly, the message from suppliers of these enabling tools is "Use our tools to aid and speed up your product innovation and management flows." None is positioned to improve a line's Outcome Velocity and Change Capacity.

Companies have made significant investments in enablers. However, these investments primarily focus on flows related to the development or refinement of a single product. The flows they enhance do not represent the complex system flows that facilitate the emergence of system parts and impactful strategic moves.

However, we're seeing constructors develop or enhance the components of a complex product line system directly with minimal human effort. This capability is a game-changer. Why? By streamlining flows and maximizing the impact of key parts, product line systems achieve greater Outcome Velocity and significantly improve the system's Change Capacity.

Notable constructors will reshape product line strategies and strategic moves. Their importance to a line's performance and Change Capacity makes them critical tools for forming and carrying out a product line's strategy. This role in strategy implies a need for teams to focus intensely on each constructor through resource dedication and governance oversight.

Big Asset Constructors (pun intended)

AI has significantly enhanced constructors' ability to contribute to product line part creation and flow streamlining. This capability is most impactful when the constructor provides a new platform-lever for a product line. Developing such a constructor is a notable strategic move. It has the potential to change system archetypes rapidly and, as discussed, lead to industry disruptions.

Tesla offers a notable example. In its pursuit of a product line for the impending $400 billion autonomous driving market, Tesla made a staggering $1 billion investment in an enabling asset: a uniquely designed computer known as DOJO. This computer was created to process vast amounts of visual data. There is no other computer like it.

The DOJO supercomputer is a billion-dollar tool designed to help the company configure a complex platform-lever algorithm. DOJO itself is neither the platform-lever nor does it alter the structure of their planned autonomous driving product line.

DOJO is just one of many tools that Tesla believes it needs. We don't know whether DOJO will succeed or become just another supercomputer. This mega-asset, along with its task-specific investment, represents a significant commitment. It has the potential to establish Tesla in the emerging autonomous driving industry.[78]

Most interestingly, Tesla also uses DOJO to build the AI core (platform-lever) of Optimus, the company's line of humanoid robots. Some experts in the field say this venture's value as much bigger than all of Tesla's other efforts. Its success and that of autonomous driving would seem impossible without DOJO. Constructors present enormous value to product lines and their parent organizations.

Constructor Proliferation

The growth of constructors has surged due to software and API integration. Consider how they can generate code, text, and images. If software or content is a key component of a product line system, there's a good chance these constructors will play a significant role as system contributors.

Over the years, major companies have developed unique programming languages to enable extensions of their platforms and pave the way for better outcomes. Google created GO and TensorFlow; META developed Hack, Apple introduced Swift, and Microsoft expanded Visual Studio. These programming languages enhance key components within each company's product line as well as many of their customers' product lines. However, these advancements have always required workflows that necessitated specialized software skills.

Constructor agents have the potential to replace workflows, improve information flows, and assist decision flows with near-instant code production. And most profoundly, constructors are likely to create code and enhance AI learning, which improves how constructors create all system parts, agents, and flows. Consider this for a moment and ask yourself how product innovation and management will evolve.

Tools and Practices as Powerful Enablers

Most marketers and engineers are familiar with the Jobs-to-be-Done (JTBD) theory, as introduced by the late Clayton Christensen.[59] Much credit must be given to Tony Ulwick and his firm, Strategyn, for laying the groundwork for JTBD theory and for operationalizing it within companies.[58] It's a powerful approach to understanding and defining customers' needs. While we know how to apply JTBD to benefit a multi-product system, most practitioners use it primarily to support their single-product focus.

The issue is that using JTBD to support single-product thinking does not help in creating Verb-based market segments within the product line stack. These crucial components of the product line arise from a cluster analysis of JTBD data and its application in identifying Verb-based market segments that enable multi-product thinking. A product line without these essential components is at a significant disadvantage.

Smart product line assemblage requires a full stack. Without a full stack, enhancing a line's competitiveness, cash flow, and customer satisfaction is a gamble at best. This means companies must invest in identifying the multi-product JTBD clustering that works best for their line. I refer to these clusters as Verb-based market segments.

JTBD segmentation is a vital enabler for establishing the multi-product context for Product Line Systems Management, rather than just its single-product application. It is essential for developing a complete product line stack.

Fighting Single-Product Thinking

Instead of focusing on the mismatch between old enablers and product line systems management, let's examine an enabler that is specific to complex systems. Dave Snowden, an expert in knowledge management and complexity theory, developed it.

Snowden calls his methodology "Estuarine Mapping."[79] It helps teams categorize constraints and plot them according to the time and effort needed to address or avoid them, as well as the potential benefits of doing so. This technique is not limited to product lines and does not use product line management terminology.

Understand Snowden's actions by applying Estuarine Mapping. He leverages a complex system's natural adaptation to navigate roadblocks, obstacles, and limitations.

Snowden noted that complex systems, unlike rivers, do not flow uniformly in a single direction. Instead, he emphasizes that a complex system can flow both forward and backward, ideally more forward than backward. He compares complex systems to the flow of an estuary, where some water moves forward, some circulates upward, and some flows backward. The flow in an estuary depends on factors such as tidal forces pushing the water and constraints like rocks and organic matter that guide the flow in various directions.

All complex systems adapt. The revelation is that complex systems adapt by following the path that requires the least energy. Snowden's Estuarine mapping leverages this principle by guiding teams to identify and define the system's constraints, helping them determine the changes necessary to overcome those constraints while expending the least energy and time.

Estuarine Mapping is a complex system enabler that combines mapping constraints with actions aimed at navigating or overcoming them. Its purpose is to enhance the system's near- and long-term outcome flow, and it is organized as an event involving a team or group of knowledgeable participants. With some adjustments and the addition of option maximization, it could be integrated into daily management practices, similar to how project management tools are utilized.

In product line systems, performance concerns extend beyond minimizing time and energy. We aim to maximize future options and avoid limiting the line's advancement; mitigating future constraints and building options enhance the line's capacity for change.

The Estuarine Mapping enabler for product line systems aims to create strategic initiatives that deliver the greatest benefits for the least effort and investment while enhancing Change Capacity. In economic terms, it would assist teams in developing an "efficient frontier" of a product line's strategic moves, where sets of strategic moves (not each move) represent a point on the "frontier" line.

Consider for a moment: How valuable is a system enabler that can create and plan the near-best strategic moves for a product line? Which product line team wouldn't desire this enabler?

Old-Best Practices

The common approach to setting up enablers has been to copy "best practices." This involves learning what has worked at other companies and mimicking it. Unfortunately, because the enablers focus on single products, this approach further reinforces single-product thinking within the organization. Although well-intentioned, the best practice implementation strategy does not effectively address efficiency and change.

A fundamental principle in PLSM is to develop capabilities and enablers based on the resources you currently possess and, whenever feasible, to avoid dismantling or discarding existing enablers. Only after addressing your current resources should teams consider implementing new processes and practices. If existing enablers do not support your product line's flows as intended, teams must establish new practices. First, modify old enablers to effectively address multi-product efficiency and the line's Change Capacity. Then, create new enablers specifically designed for your complex product line system to support everything that needs assistance.

Key PLSM Principle

Effectively managing your product line as a complex system empowers teams to transform their offerings. However, transformations and significant changes only succeed when paired with operational efficiency. This necessitates a rigorous application of orderly system methods such as Lean and the Theory of Constraints. These practices intentionally eliminate unnecessary costs, enhance output, focus on non-outcome-related flows, and remove barriers to achieve greater gains. This orderly system work is crucial for substantial product line changes.

Notice how the two types of systems, orderly and complex, complement each other in terms of their orderly and dispositional behaviors when applied to the relevant system substrate. The key is to identify orderly systems within the complex product line system.

Suppose your team is making a strategic move. Practices such as lean and TOC often enhance the effectiveness of this move during execution. Product line teams should prioritize orderly system practices over dispositional approaches when planning these moves. If an undesirable outcome arises from the move, the focus may need to shift to orderly system practices to ensure the move has a positive impact.

Consider Elon Musk's experience with Tesla. Walter Isaacson described this in his 2023 biography of Musk.[80] He details how Musk and his teams made significant strides in developing a new modular approach to building the Model 3 in Fremont, California. This project aimed to launch Tesla's first mass-market vehicle.

One week after launching the vehicle at its mid-range price, Tesla received down payments on over 300,000 orders. The key to success was complementing its unique modular design (platform-lever) with a distinctive modular production process (also a new platform-lever). Their concept involved using robots to automate everything. However, the company faced significant production throughput issues, having created an automated process filled with bottlenecks. The situation was shaping up to be a product line disaster.

Musk recognized that the company would fail if his team did not address the bottleneck issues. His response was to confront the problem with relentless focus. He moved into the factory in 2017 to work and observe the production lines around the clock. He was determined to resolve the issues. And that's what he accomplished, averting the company's collapse and enabling full production in 2018.

Musk's Algorithm

Musk's key lesson from the experience was that companies must initiate a focused and intense push for efficiency after making significant changes. He summarized the efficiency drive in a five-part "algorithm." It's a causal-efficiency push to make major new approaches and strategic moves effective.

1. Question every "requirement"

2. Delete any part of the process you can

3. Simplify and optimize everything

4. Accelerate the cycle time

5. Automate the flow

The algorithm resonates with experts in Lean and the Theory of Constraints, imparting a hard-won lesson to innovators and change agents. A key PLSM principle encapsulates this perspective:

Disciplined efficiency methods, like root cause analysis, should complement strategic moves derived from complex systems. Consider using orderly system tools, such as the

Theory of Constraints, to streamline the flows resulting from complex system moves.

Focus on capturing the fundamental truths—the basic issues, problems, or challenges that may impede the intended benefits of any strategic move. Then, reason up from there rather than reasoning backward with analogies. In complex systems, "fundamental truths" will likely involve multiple agents; there is rarely a single-agent truth. Plus, teams must recognize that fundamental truths may align with difficult solutions and that, if neglected, could lead the systems into chaos.

Insights, Actions, and Outcome Coherence

A roadmap diagram for a product line visually communicates how the product line system is intended to function. It outlines the key components of a product line and highlights the goals that teams strive to achieve with technologies, markets, products, and platform-levers. However, the diagram does not clarify why or provide a comprehensive view of the associated risks and opportunities.

A system enabler to facilitate roadmapping and, more importantly, bring insights, the "magic sauce of strategy," into play with system risks and opportunities. Roadmapping requires a graphing enabler, while the insight to risk and opportunity rationalization necessitates a narrative enabler. Teams need both graphs and narratives. Creating and improving this combined enabler is a significant task in product line systems management.

Machine Learning and Artificial Intelligence

System enablers play a significant role in advancing product line systems. The question facing teams is whether their enablers focus on improving one product at a time or managing the entire product line. This is an important issue to consider. Nowadays, you often hear about how AI can perform a job better. This isn't about enabling a task that couldn't be done; it's about doing the current job more effectively.

Product line teams do not need to address their outdated single-product thinking. Instead, they must shift to a multi-product approach and recognize its complex systemic nature. While some individuals working on product lines desire this transformation, it is the responsibility of upper management to facilitate it.

A governance body must oversee the product line. It is a critical entity necessary to ensure that the product line performs efficiently and can adapt. Without effective governance, a product line's ability to adapt and thrive in the long term will diminish. I will discuss governance further in Chapter 23.

Assemblage Map of Enabler Changes

A product line team's approach to enhancing its enablers is as important as the strategy established for the line. Consider the core strategy for the product line's components and architecture to boost the line's Outcome Velocity. The enabler strategy focuses on improving the line's efficiency and adaptability.

The roadmap for enablers, called the Assemblage Map, is as important as the roadmap for a product line's key components. Enablers should not be considered afterthoughts introduced to a company by a consultant or system vendor after the key components are established or altered. Instead, teams must create coherence among enablers and throughout the system's core components.

A primary concern for product line teams is that enablers are crucial for integrating Change Capacity into their systems. In the

Accelerating Change Race, teams will realize that postponing enabler adjustments until after a strategic move can be problematic. Moreover, it may prove fatal if the move involves changing a system's archetype.

I share examples of Assemblage Maps that illustrate the advancements of enablers in our online appendix. However, please understand that there is no standard for such maps, but I provide two simple guidelines for creating them:

1. Ensure they communicate the current and future state of all recognizable enablers.

2. Use clear narratives to explain how the changes to an enabler will enhance the product line's Outcome Velocity or improve the system's capacity to change.

Enablers, Constructors, and The Accelerating Change Race

Keeping pace in the Accelerating Change Race requires teams to reinvent old enablers and develop new ones. It also involves creating constructors that facilitate the construction of system flows and components. Outdated practices and tools will not meet the demand for greater efficiency and the simultaneous changes that must occur.

The contemporary need for efficiency and change must encompass system enablers and builders. The good news is that AI, along with some powerful processing capabilities, serves as a significant aid in building enablers and constructors. The resultant support has become as essential as the key components of the line.

Chapter 21 Summary

Enablers & Constructors

- **System Enablers:** These system agents are practices, processes, environments, tools, and assets that help or promote a product line's workflows, decision flows, information flows, and outcome flows. They aid the product line's performance.

- **Unaffected Archetype:** Enablers are not core parts within the system and don't change the system's archetype. They can be temporary or permanent.

- **Purposeful Use:** Teams and governance groups assume responsibility for identifying and implementing enablers for their product line.

- **Single vs. Multi-Product Orientation:** Enablers have limitations, especially if they focus on single-product efficiency rather than multi-product change capability. Old-school enablers designed for orderly systems don't perform well with complex systems.

- **Complex System Enablers:** Enablers like Estuarine mapping are created for complex systems. They can help teams identify and address constraints to improve a product line's Outcome Velocity.

- **Asset-type Enablers:** Companies invest in asset-type enablers like Tesla's DOJO computer to gain a competitive advantage in creating system parts and agents.

- **Product Line Part Roadmaps:** This new enabler focuses on helping teams translate insights into actions and decisions that improve a product line's Outcome Velocity.

- **Product Line Enabler Roadmaps:** A roadmap for the advancement and expected change of system enablers is a major part of product line systems management. It lays out a coherent plan for gaining value from enablers.

Chapter 22

Product Line Change Theory

Making Change Doable

The Theory:

Product line systems change more rapidly and efficiently when requisite changes to enablers and ecosystems, at their finest level of granularity, accompany key part changes.

The Product Line Change Theory suggests that you must change the system's assemblage to achieve the greatest performance gains from modifications to the product line's key part stack, including scaling any part of the stack. Furthermore, these changes must be at the most granular level of enablers and supporting ecosystems.

The theory has significant implications for the Accelerating Change Race and for integrating change into core product offerings. It offers a critical perspective on what is needed to facilitate substantial change in core product lines.

Changing the System

When managers view product offerings solely through the lens of single-product, orderly systems, changes at the most fundamental level seem illogical. No agents exist in this context. This explains why individuals working in this traditional environment often assume their simplest option is to claim that core products cannot change, a belief that has persisted for decades.

However, significant changes to product offerings can occur when teams and their leadership understand the importance of updating and modifying the agents and system assembly of the product line, along with adjustments to the product line stack.

By recognizing and managing a product line as a complex, multi-product system, we can implement significantly greater and more impactful changes to the entire product offering. This approach requires that we understand the system at its most fundamental or granular level. (See Chapter 2)

Roadmapping and Assemblage Mapping

Managing change in the relationships between key parts and the system's other agents is central to PLSM. The connection of key parts and the system's supporting agents demands a unified approach between product line roadmapping and system assemblage mapping. One focuses on the dynamic movement of the line's key part stack, and the other on the progression and modification of system enablers, constructors, and ecosystems. Their interplay and alignment determine the system's capacity to perform and transform.

Any change to a line's key part stack represents a strategic move. When a move involves a market pivot or an archetype transformation, changes to enablers and ecosystems will be more significant than those for incremental changes to key parts. Consider how system archetype changes (platform-lever type changes) will demand changes to a business model and the corresponding chain-link alignment, i.e., the line's internal ecosystem.

The theory also suggests that external ecosystems, such as supply chains and sales channels, must change in alignment with changes to the line's key parts. Anyone who has experienced a Noun-based or Verb-based market segment pivot, or an addition of a technology building block, is likely to have encountered problems when adjustments to external ecosystems were not made promptly.

The key principle that the theory underscores is that as changes to key parts are being conceptualized, teams must also plan for and prepare to invest in modifications to the system's other agents at the most detailed level.

The Old-School Change Issue

Traditional product innovation and management often struggle to enable and support radical innovations, i.e., significant changes to key parts. The results tend to be subpar because traditional managers don't recognize that they are operating within a complex multi-product system, where the complexities of this system take precedence. However, the issue is that a lack of harmony between key parts and the support agents negatively affects the emergence of outcomes.

The key issue with the traditional single-product, orderly systems approach is that mismatches between strategic moves and system agents can accumulate, undermining the performance of the product line. If these mismatches are not addressed, they will grow, leading to system problems that can spiral out of control.

Real-World Change

Consider the changes currently happening in the automotive world. The electric automobile industry witnessed a new system archetype disruptor with the emergence of an electric vehicle powertrain platform-lever (see Chapter 12). Every internal combustion engine (ICE) product line became at risk, prompting their parent organizations (Volkswagen, Ford, GM, etc.) to launch their electric vehicle (EV). Unfortunately, they did so without adequately addressing the necessary changes to their product line's agents and ecosystems.

Most incumbents sought to integrate their EV offerings into existing product line systems. This strategy resulted in conflicts between the new EV components and the established system agents. Not only did it stall their EV initiatives, but it also had a negative impact on their ICE products.

Industry dynamics do not bode well for the incumbent companies. While these automakers were content to watch EV demand in the U.S. decline, global demand is on the rise. Dedicated electric vehicle (EV) companies in China have grown their market share of the entire automotive industry from 2% in 2010 to 30% by 2025. This growth, combined with low wages and government support, gives companies like Xpeng and BYD significant advantages over Western manufacturers.

The situation worsens for Western automotive product lines as we approach autonomous driving, the next disruption in the industry. The archetype and key part changes, along with the corresponding changes to enablers and ecosystems, will be significant.

The Theory's Basis

All strategic moves aim to enhance the speed of a line's outcomes. Since they also involve changes to the system's key parts, they will depend on and be sensitive to the Product Line Change Theory.

Please don't view the theory as a brilliant discovery by the author. Instead, the theory adapts the work of complexity experts to product line systems. Dave Snowden deserves credit for the insights that inform the theory. Snowden articulates these insights in his Estuarine Theory, which suggests that complex systems change more like an estuary between a body of water and solid ground. He explains that an estuary is a strong analogy for complex systems.[56]

Snowden's thinking extends beyond product lines, which is different from mine. He focuses on all complex systems. The analogy is that just as tides push water into an estuary (the system), change occurs naturally based on its constraints, forces, and constructors (see Chapter 21). Interestingly, these components also determine the system's substrate and its behavioral patterns. As I share in Appendix 2, our substrate is the product line system.

Snowden's contribution is twofold. First, he points out that every component of a complex system has an energy gradient and a duration related to change. Second, he demonstrates that identifying each component's lowest energy and time path to change offers a method that is far superior to trying to push the entire system. The most effective strategy, Snowden argues, is to alter the relevant system components at the substrate's lowest level of granularity.

Product line teams can adjust their system's components, agents, and constraints by influencing, nudging, reforming, eliminating, and creating. Conversely, attempting to change everything related to product offerings as a whole is problematic and often counterproductive.

Exploiting The Change Theory

The Product Line Change Theory explains why traditional approaches struggle with implementing change and provides an approach for introducing transformation into product lines. However, as we seek to leverage the theory for the benefit of the line, we encounter four critical questions.

1. How should we view and best understand potential product line stack changes?

In Chapter 5, I explore the key part stack and how a multi-product stack differs from a single-product stack. Chapter 9 delves into a system's assemblage. Table 9.1 outlines the potential key parts, which include technology building blocks, products, platform-levers, and market segments. A complete assembly, however, must also include system enablers and ecosystems. The theory suggests that when key parts change through addition, subtraction, or modification, we must deliberately adjust the supporting enablers and ecosystems to achieve the desired performance.

2. How can we determine what enablers and ecosystems must change and how they should change?

First, identify and map the enablers, constructors, and ecosystems that have already been established to support and drive your product line system. These processes, templates, practices, and entities facilitate the various flows within the system. Analyzing the potential system flows (work, information, decisions, and outcomes) required after making product line stack changes, compared to what is likely with current enablers and ecosystems, will reveal misalignment and potential bottlenecks. System assembly corrections must then be crafted to offset these problems.

The goal is to have the correct system enablers and ecosystems ready to go when the key part change (strategic move) launches.

It is wiser to identify and plan changes to agents and ecosystems simultaneously with the development of the key part. To do this, teams should carefully consider and model the system's changes. They accomplish this through workshops focused on unifying the line's roadmap and assemblage map. This work should take place in the early stages of the key part's development, well before the launch or roll-out.

Changes to the agents and ecosystems must also be communicated to all influencers and actors as soon as they are identified. Influencers and contributors should then evaluate, analyze, and strategize the best approach to implement the necessary changes. The primary concern is that delays and failures in agent changes can affect other parts and agents of the system, thereby impeding the speed of the line's outcomes.

3. Can we add temporary agents to compensate for the shortcomings of existing agents?

Yes, temporary agents are often necessary to achieve the desired performance from changes in key areas. The larger the change, the more likely it is that temporary agents will be required. These agents may include both orderly system methods, such as the Theory of Constraints or other RCA methods, as well as complex system methods such as constraint mapping and force field analysis.

You'll often see ad hoc teams formed to do everything necessary to ensure the success of a strategic move. Temporary agents take this a step further. In complexity theory, Snowden refers to these as "scaffolding." They act as interim agents that enhance the system's performance while other, more permanent agents are being developed.

4. How do we lessen the task of changing key parts and agents concurrently?

When planning a strategic move—the change of one or more key parts—teams should explore and understand the path of least resistance for the transition. This can be done through "Agent Change Workouts" and performed using Estuarine Mapping, which was introduced in the last chapter. This workout should occur concurrently with conceptualizing the key part change.

In this application of estuarine mapping, teams view the mismatch between an agent and the system's potential flows as a constraint. Estuarine mapping then directs the team in exploring the energy and time required to modify each flow-constraining agent, or, in some cases, to create new agents. From this analysis, a team should gain sufficient insights to chart the best path forward.

Change Capacity Matters

It is important to recognize that all product lines have some capacity for adaptation; at times, they may show significant capacity, while at other times, they may display much less.

However, developing Change Capacity does not require an immediate change to the system's key parts. It is both possible and beneficial to enhance Change Capacity within a line before planning any changes to the product line stack.

Major problems arise when the system's Change Capacity is low and key parts change nonetheless. In these situations, teams may be forced to impose changes on the supporting and hindering agents. Unfortunately, this reactive approach is much less likely to produce positive results. You'll find more information on building Change Capacity in Appendix 5.

Corporate Innovation and the Change Theory

Perhaps the most significant self-inflicted issue in corporate innovation is that these initiatives lack guidance from the principles outlined in the Product Line Change Theory. Corporate innovation efforts focus on creating key parts in a product line stack while neglecting the creation or reform of system enablers and ecosystems. This issue stems from the single-product, orderly system paradigm in which they operate. The theory and its role within a product line's complex, multi-product system shed light on the reasons behind many corporate innovation failures.

Many managers in corporate innovation intuitively recognize the truth about system agents and ecosystems, yet they often feel powerless to address it. Consequently, they focus on creating key components and their stacks, overlooking the necessity to develop or modify system agents and ecosystems. This is a mistake.

The Product Line Change Theory offers a necessary reexamination of corporate innovation's focus, objectives, and methods for achieving those goals. It advocates for redirecting core business transformation to enhance the Change Capacity of system agents and ecosystems. This recognition begins with viewing product lines as complex, multi-product systems. I elaborate on this in greater detail in Chapters 27 and 28.

Business Transformations and Product Line Change

While corporate innovation is underway, core businesses often invest considerable effort into organizational and digital transformation. However, this transformation mainly aims to enhance efficiency within the agents and ecosystems that support an existing product line.

These common transformations do not include product line transformation; in fact, they assume product lines will stay constant. Consequently, these well-intentioned efforts often diminish the capacity for change within the existing product line systems. It's no surprise, then, that corporate innovation initiatives struggle to find direction.

Strategic Moves

In Chapter 15, I examined how to create a coherent stream of strategic moves to enhance a product line's Outcome Velocity— positive gains in cash flow, customer satisfaction, and competitiveness. Each strategic move represents actions that alter the line's key components, whether focusing on improvements in short-term or long-term Outcome Velocity. A strategic move consists of one or more actions and is greatly influenced by its design and implementation path.

In product line management, teams develop potential strategic moves focused on improving their line's Outcome Velocity and addressing constraints that may arise. The team then assesses the potential strategic move for coherence and its compounding gains with other moves, considering the energy and time required to adjust the system parts and agents. This approach helps teams understand what they're getting into before executing the move. It's an analysis from market, technology, and system perspectives.

Figure 22.1 below illustrates a process for creating and implementing strategic moves. Notice how this flow begins with a charter to guide the development of the move. Such charters originate from product line roadmapping and systems assemblage mapping.

As a principle, I don't recommend moving directly to executing a strategic move, especially if it was developed outside the system. Instead, translate these potential actions into charters that can be integrated into the roadmap and lead to executable strategic moves.

The flow directs teams to create a move that aligns with the charter, revise the strategic move based on coherence with preceding and subsequent moves, and pursue compounding gains. The flow then transitions into fine-tuning the move according to necessary system changes at the lowest granular level. Execution plans that guide technology, market, offering, and system changes can then be developed. If necessary, the charter can be revised, and the flowchart can be updated accordingly. Nonetheless, the flow always starts with determining whether the charter is beneficial to the product line's key part roadmap or the system assemblage map.

Execute strategic move when ready.

Create or Re-Create **Strategic Move Charter:** Purpose, Direction, and Potential Constraints

Create an **Execution Plan** and System Assemblage Change Plan, followed by Orderly (RCA) Outcome Flow Unblocking

Create strategic move based on improved outcome velocity and/or change capacity.

Refine strategic move based on change energy and time. Map the system change energy and time at its lowest granularity

Recast strategic move based on coherence with other moves, compounding gains, and system assemblage map.

Figure 22.1 *A workflow for creating and refining product line strategic moves.*

The concept of system Change Capacity also aids in understanding the system's current ability to change, indicating its readiness to adapt to various key part change scenarios.

Successful Change to Core Offerings

It's interesting to explore instances where changes in support agents must accompany modifications to a line's key components. These parallel activities typically remain private to a company, except when things go poorly. That's when shareholders and the business press are quick to highlight the issues and management's shortcomings in handling the situation. The challenge within the company is that sometimes the change needed in enablers and ecosystem is apparent, while at other times it is not.

Consider Tesla's platform-leveraged change, as described in Walter Isaacson's 2023 book, Elon Musk.[76] According to Isaacson, in 2018, Tesla had just launched its Model 3, and automated production was significantly underperforming. The company and the product line would be in jeopardy if it didn't get fixed quickly.

Because the Tesla team did not figure out the changes before they were needed, an all-hands-on-deck situation arose. Musk famously slept on the shop floor while focusing on the bottleneck issue. In effect, the Tesla team created a temporary production flow agent to overcome the constraint.

Scaling AI

A crucial aspect of product line management is determining how to effectively scale a product line stack. Scaling always represents a significant change to a system. When AI is involved, the change is particularly demanding because we have never had to deal with such a massive and impactful technology that is changing so rapidly.

To implement AI-driven system changes, teams must first determine whether the AI instance acts as a key part (a product, a platform-lever, or a technology building block) or as a supporting agent, such as a flow enabler or a constructor. This determination will guide the scaling efforts in various directions.

The challenging aspect of scaling AI is that its behavior and the behavior it induces on the system can vary at different scales, even when considered independently from the whole. AI is a complex system (a sub-system of the product line system) with behavior that is not entirely predictable. If AI is deployed as part of the product line stack and is integral to the customer experience, then new system agents must monitor and respond to potential adverse behaviors.

In Chapter 12, I discussed how AI will alter the platform-lever mix for many product lines. When this occurs, the archetype of these product line systems will transform, necessitating significant adjustments to the parent business model, the chain-link interplay, and the system agents and ecosystems.

If the AI application is implemented to support a non-key part agent, the desired outcome may focus on reducing costs, enhancing competitiveness or increasing customer satisfaction. However, these agent applications will not be all-at-once bespoke solutions, but rather will need to evolve.

If some cases, AI scaling may remove a system constraint. If it does, the first question should be, 'What's the next constraint that can be negated or mitigated, and can an AI agent help?' Remember, there is always a "next constraint." Your task is to identify what it might be, spot it as soon as possible, and mitigate it quickly.

Assemblage Mapping and Agent Change Portfolios

In traditional product innovation and management, companies commonly use product roadmapping and project portfolio management; one emphasizes the strategy or direction of products, while the other concentrates on resource allocation to achieve desired outcomes. PLSM, in response to change theory, must take this a step further. It requires strategic planning and the implementation of system changes, both to the product line stack and to the agents and ecosystems that support it.

In PLSM, a product line roadmap communicates our direction for the components of the product line stack. The Assemblage Map illustrates the adjustments necessary for all system agents and ecosystems to enable a successful product line roadmap. A

Project Portfolio includes all projects related to the product line stack. At the same time, the System Change Portfolio offers an overview of agents and ecosystems being created or modified, along with the resources required for success. Currently, there are no established or traditional "best practices" for these significant contributors.

You will find much more information about assemblage maps in the online appendix, along with evidence—both factual and counterfactual—that supports the theory.

Product Line Change Theory and The Accelerating Change Race

The Product Line Change Theory is essential for keeping pace in the Accelerating Change Race. It provides guidelines for adapting a product line system to stay aligned with rapid changes. Traditional methods encounter significant challenges in this area.

When a key part of a product line stack changes, supporting agents must also adapt—determining what they are and how they need to evolve is a necessary and fundamental skill in product line management.

Chapter 22 Summary

Product Line Change Theory

- **Challenges of System Change**: Traditional product management struggles to enable radical changes. By managing product lines as complex systems rather than a set of individual products, companies gain a viable approach to changing core product offerings.

- **Theory Principles:** Changes should target the system's lowest granular levels of components (specific constraints, parts, or agents) for faster, more effective transformation. This approach, inspired by Dave Snowden's Estuarine Theory, emphasizes leveraging the energy gradients and time paths of individual system components.

- **Traditional Approaches Fail:** Single-product management methods cannot effectively exploit this theory because they do not recognize the complex system components.

- **Strategic Moves:** Strategic moves aim to enhance product line Outcome Velocity—its cash flow, customer satisfaction, and competitiveness. Planning strategic moves must consider the constraints, coherence, and potential compounding benefits of changes within the system.

- **Workflow for Strategic Moves:** The chapter also shares a workflow for strategic moves that includes both key parts and system agent changes.

Chapter 23

PLSM Governance

Enabling and Clearing Paths

L et me share a story before discussing PLSM governance. I'm reminded of it every time the topic of governance comes up.

Governing Innovation

Several decades ago, I worked with a client team at Exxon Chemical that embraced intrapreneurship. For them, the intrapreneurial approach appeared ideal for developing and commercializing an exciting new technology. My role as a consultant was to assist the company in advancing a new catalyst technology developed by its R&D department.

The metallocene catalyst acted like alchemy in a reactor, creating uniquely structured polyolefins. These polymers, including polyethylene and polypropylene, represent a significant market. The polyolefin product line was a key contributor to the business unit's earnings.

The uniqueness of the new catalyst-produced plastic meant that the product's value would be significantly higher than what Exxon was experiencing with its more commodity-like polyolefins. It also suggested that the volume would be considerably lower.

As a young engineer, I was in awe of the scale of Exxon's reactors. However, I didn't realize the challenges that came with the company's production scale. When operating a reactor, plastic is produced rapidly, and shutting it down is undesirable because it leads to increased costs and waste. The challenge for the intrapreneurial team was that introducing their catalyst into a large reactor would yield far too much of the new plastic, and they couldn't simply swap out the catalyst or turn the operation on and off. The costs would be astronomical. But here's the major issue: this new material had not yet penetrated any market. There was no demand.

Intrapreneurs at Work

The ingenious intrapreneurs took it upon themselves to solve the problem. They began production by using the plastic made in the company's polyolefin pilot reactor, which was only a fraction of the size of the full-scale reactor.

Typically, the pilot reactor was used to test polymer variants requested by customers. This innovative solution was precisely what intrapreneurs were encouraged to pursue. Gifford Pinchot[9], the creator of intrapreneurship, would have taken pride in this.

Unfortunately, the clever workaround was completely undermined during a presentation to management by the team's leader, its top intrapreneur. I'll call him John. John presented their issues and solutions to the venture's governing board, a team of Exxon Chemical executives.

Poor John. These brilliant executives teamed up against him and called him an idiot for interfering with the core business. They viewed the test reactor as their bread and butter. It was used to test small variations of existing polymers and served as a tool for closing sales.

What We Learned

John's fate illustrates why intrapreneurs are labeled as "entrepreneurs-in-training" rather than entrepreneurs within companies. A few days after the presentation, John was "encouraged" to leave Exxon. He complied and, along with several colleagues, founded a new company that had no ties to plastics or metallocene catalysts. He became a true entrepreneur.

This case forces us to examine several issues.

First, the executive team worked against the venture rather than with it. Why did that happen, and how can it be prevented?

Secondly, the metallocene opportunity was perceived as an independent venture rather than a growth opportunity for the core polyolefin product line. At that time, no one recognized that the catalyst-reactor combination represented a new platform-lever that needed to be integrated into the polyolefin line. Was it the best course of action to separate the venture from the core business? Why didn't the core business pursue the commercialization of this opportunity?

The third issue we still encounter today is that intrapreneurship rarely, if ever, produces meaningful or impactful results. It consistently conflicts with core business objectives. That is what intrapreneurs are trained to do—act contrary to a core business. Intrapreneurship is not a viable solution for addressing the efficiency-change dilemma. A modern-day intrapreneurship guru told me it will work today because the millennial generation differs from the older baby boomers. To put it politely, that's a bunch of nonsense. But that leaves us with a critical question: Who's going to stop it?

Each issue is a governance issue.

Guiding Innovation

Governance has historically served as a parental-like control mechanism, which is not what most innovators or product developers seek. The origin of the term "governance" is rooted in IT departments, not innovation. Its transition into the realm of innovation occurred alongside project portfolio management, which evolved from a strong focus on project management. A governance body did not oversee creative thinking, marketing, or technology management; instead, the executive team took on this role but never accepted responsibility for supporting and empowering product development and innovation. These top managers maintained oversight at a company-wide level.

Innovation had already gained significant momentum in engaging senior management and top leadership. This was the cornerstone of Mike McGrath's (the M in PRTM consulting) PACE process[81] in the late 1990s. Robert Cooper's Stage-Gate[82] work highlighted the importance of involving top managers in decision-making. These contributions occurred well before the Agile movement impacted companies in the early 2000s and before project portfolio management gained traction a few years later.

The common perception of governance is that it should create trust, facilitate transparency, and encourage a balance between short-term and long-term decisions. However, I see it differently. PLSM governance should reflect an organization's culture and inspire a product line team to engage with and lead their organization in all matters related to their product line. Trust and transparency must arise from organizational behavior and not be limited to a product line. Trust and transparency are issues that affect the entire business, not just a product line.

Many people also view governance boards as entities that keep separate interests "in balance." My experience suggests that "balance" may seem appealing, but is ultimately meaningless. When an expert or manager in portfolio management or strategy brings up "balance," my nonsense flag goes up. They're just making things up. They can't define what balance is. Yet, who would argue against it? Have you ever heard anyone say they don't want balance?

In PLSM, the issue isn't balancing the portfolio; it's about improving the product line's Outcome Velocity. The Governance team's role is to empower the product line team to enhance the line's Outcome Velocity, rather than creating a meaningless "balance" between a near-term and long-term vision.

My friend Bebee Nelson and her colleague, the late IMD Professor Jean-Philippe Deschamp, published the first book on Innovation Governance in 2014[83]. Their insights highlight the essential leadership and guidance roles needed for innovation. They emphasize that the primary mission is to influence the "how" of innovation, not the "what." It's about enhancing the organization's capacity and capability to innovate. A simple analogy I use is that the Governance team must ensure the health and well-being of the goose, which must then concentrate on delivering valuable golden eggs. Governance nurtures the goose; teams produce the eggs.

Major Additions

I urge all executives and managers to differentiate governance from decision-making. These are two distinct concepts. Governance involves overseeing the management of the product line system, while decision-making encompasses all choices and judgments within the system. PLSM governance should guide and enhance the system's workflows and part creation, rather than make decisions for the product line team.

PLSM Governance must facilitate the connection between the product line and the entire business, and vice versa. This requires that individuals in governance roles have a deep understanding and knowledge of the product line's archetype and strategy, the business unit's strategy, its business model, and the chain-link strategies.

A significant part of managing a product line involves thoroughly understanding the system's components and constraints. Governance teams embrace this understanding.

Don't Control (unless truly needed)

The job goes beyond limiting, controlling, and guiding; its purpose is to empower and enable product line systems management. This distinction is what makes PLSM Governance a system enabler. For most product line systems, governance is the most critical enabler.

Insufficient governance can undermine a product line. Without effective governance, the strategy and coherence of decisions and actions may falter. However, governance alone cannot guarantee product line outcomes or increase Outcome Velocity. Rather, it ensures that pathways to desirable outcomes remain accessible.

So, what should governance emphasize? What responsibilities should a governance body assume? And what are the product line teams expected to accomplish without enabling governance? These questions are critical.

Communicate and Focus

The governance body should consist of senior managers capable of influencing critical chain-link functions. They are responsible for guiding the product line team and helping the team work across chain-links.

Strong communication requires an openness between the product team and governance members, and it can only happen when trust exists across all parties. This fundamental principle is often overlooked. It appears that this topic only arises in the absence of trust.

Oversight Feedback

Governance establishes constraints while allowing boundaries to be flexible. It also offers feedback on potential investment decisions related to large-scale components, new players, and impactful enablers.

Granted, some fixed boundaries are necessary, like "Don't kill the customer. Don't kill the company. And don't send us to jail." However, governance constraints can hinder system advancement and change, especially when guiding boundaries turn into controlling limits. You know this is happening when you hear things like "the customer is always right," "cash is king, all else is irrelevant," and "always Cover Your Ass, but remember my ass is more important than your ass." Governance should prevent such forced limits unless critically necessary.

Not all customers, especially large clients in B2B markets, are always right. That's a customer service mantra. It's not a mantra for product innovation and management. And while cash is important, it should never undermine the significance of customer satisfaction and competitiveness. The Outcome Velocity vector challenges this notion by emphasizing the need for continuous improvement in all three areas—cash flow, customer satisfaction, and competitiveness in both the near and long term. The governance team must adopt the Outcome Velocity vector as the primary performance indicator for the line.

You'll also notice that CYA is a common and potentially dangerous behavior in large organizations. It hinders insight generation, and if you kill insight generation, you also negate the fuel that drives change. In complex systems, Governance teams must embrace the uncertainty that product line teams must operate within. To do otherwise, as CYA does, is to double down on efficiency and squash any chance for change.

Enablement

Enabling governance acts as the linchpin between business strategy and the performance of the product line. As discussed in Chapter 15, the product line strategy answers Roger Martin's key strategy questions: "Where to Play and How to Win."

Consider how effectively this crucial connection benefited NVIDIA. Then, reflect on how disastrous it was for Kodak.

Jensen Huang, NVIDIA's CEO, is recognized for his long-term thinking, which he incorporates into the company's strategy as "where to play" and "how to win." Initially, the company maintained a sharp focus on data-intensive graphic processing chips, setting its core product line apart from all its competitors.

The company's transition to AI processing chips took ten years to develop. The initial step involved building world-class competencies within their product line stack. This initiative unfolded through the company's Center of Excellence and is evident in the foundational level of the stack (Figure 9.1 in Chapter 9). Importantly, their shift in market orientation began long before the stock market took notice. From a systems perspective, it started by observing and focusing on the lead user applications of their GPU. Lead users, as we know, are rarely the largest customers. For NVIDIA, lead users were not found in the gaming market.

From a PLSM perspective, Huang guided his team, focused on a specific product line, in orienting itself toward the future. The focus shifted to Noun-based segments, developing a powerful "design-production" platform-lever that facilitated the delivery of Verb-based attributes. This design then extended functionality across the AI market in ways that other chip manufacturers could not. In 2014, NVIDIA had less than one-tenth the market capitalization of its major competitor, Intel. "By 2024, it surpassed that by more than tenfold, and outgrowing Intel's valuation by one hundred-fold."

I credit the success of the chip maker to the guidance and path-clearing provided by NVIDIA's governance team. As with NVIDIA, product line governance acts as an essential link between product line management and business strategy.

Then The Opposite

Kodak took a different path in the 1990s. In the decades leading up to its downfall, Kodak was a haven for innovators, attracting some of the brightest minds. However, digital photography caused the company's Noun-based market segments to deteriorate rapidly. The company was so entrenched in its old practices, assumptions, and thought processes that the fact it originated digital photography meant little.

The company's late entry into digital photography and its rigid, old-school practices focused on 35mm film left its leaders with little flexibility. They were compelled to meet Wall Street's investor-driven demands for greater near-term cash flow. Unfortunately, generating positive cash flow while top-line revenue is declining presents challenges. From a product line perspective, Kodak had minimal capacity to change. They then depleted what little Change Capacity remained to achieve better near-term cash flow.

It's important to understand that Kodak recognized the need for change years before its downfall. One rarely told story today is that the company's Corporate Innovation efforts shifted it into clinical diagnostics. Although this venture never generated a profit, it was sold to Johnson & Johnson for $2 billion in today's value. However, this sale was insufficient to address Kodak's core business issues.

J&J ultimately shut down the venture because its archetype grossly mismatched the business model of its Clinical Diagnostic division. They couldn't make it profitable, regardless of the guidance they provided to the venture. I know this story because J&J engaged me to help integrate Kodak's product innovation and management practices with theirs. I had previously consulted with both firms, and someone at J&J thought I might be able to assist. Unfortunately, I had no knowledge of the subject matter in this book at that time.

At the time, no one understood that the challenge involved a complex system. However, it's now easy to identify both Kodak's and J&J's shortcomings. Neither company had governance that was fully aware of what was truly happening. Everyone relied on the C-suite to make decisions and guide all efforts. Both the Kodak and J&J teams lacked an understanding of the effects of changing system archetypes, business models, or aligning ecosystems.

Training and Involvement

I have two pieces of advice for potential members of an innovation and product management governance team. First, familiarize yourself with product line systems management. Second, understand your business unit's strategy, business model, and functional chain-link strategies, as well as their interplay.

PLSM establishes a unique framework for thinking. It differs from the single-product context that most potential governance members are familiar with. If governance team members do not understand the multi-product, complex system context that PLSM introduces, their contributions may become counterproductive.

Understanding PLSM is essential for continuous improvement in product line performance and Change Capacity. If managers, especially the organization's top managers, apply the single-product orderly systems context to product line system dynamics, they are more likely to harm the system than to help it.

How a product line integrates with a business is just as important as the line's innerworkings. It is the governance team's responsibility to ensure that this connection remains strong. This means that the business unit's strategy, business model, and chain-link strategies must align with the product line system, and vice versa. To achieve this alignment, the governance team must understand and communicate a shared understanding of each.

Hopefully, these strategic elements work together. If they don't, coherence and alignment will be lost. It is the governance team's responsibility to realign the strategic elements. Then, it is the product line and functional teams' responsibility to reestablish flow resonance and synergistic boosts.

Managing Up to Avoid The Big Mistake

PLSM governance operates between the product line and the C-suite, which comprises the chief officers of the company. While many experts assert that the primary role of the governance board is to protect the product line team from upper management, a more critical responsibility is to guide the C-suite to prevent significant errors. If these high-level mistakes accumulate, they can harm both the product line and the overall business. However, these issues generally affect the product line first.

Three such mistakes are more common than anyone would like to admit.

1. Success hubris

This issue is common among governance teams in many large companies, particularly industry leaders. It arises from the excessive pride and attitudes of leaders shaped by their past accomplishments. Senior managers adopt the mindset that "I can do nothing wrong. My past success confirms that." However, this creates a false sense of invulnerability. I dedicate a case study to the dangers of success hubris and its devastating consequences. (See Online Appendix Case Study 1 on Keurig Coffee). .

2. Poor enablement of sense-making and insight generation

If there is little or no support for insight generation and sense-making, a product line team will likely avoid attempting to navigate the struggle between efficiency and change. Addressing rapid change and the efficiency-change dilemma with inadequate insight generation and sense-making is akin to playing a sport with both arms tied behind your back and your legs in shackles.

3. Being oblivious to uncertainty's upside and downside

The essence of managing complex systems revolves around addressing uncertainty. It is detrimental if the leaders in your organization do not recognize the importance of this.

The governance team is responsible for addressing these issues, ideally before they negatively impact the system. Their role involves highlighting the risks of not advancing core product lines and advocating for investments in insight generation and sensemaking. They must also ensure that all C-suite leaders understand PLSM and its importance in maintaining competitiveness in the Accelerating Change Race. Occasionally, governance teams may need to attract growth-oriented investors. However, this pitch is usually more effective from the trenches of product line management than from the lofty realm that corporate innovation tends to portray.

Guiding Governance

Here are a few pointers for elevating a governing body's contributions.

- Remember Agency theory, which states that subordinate actors serve as agents for superior actors and that superior actors can influence the actions of subordinate agents. The governors are subordinate agents to their CEO, and the CEO is subordinate to the shareholders. If shareholders don't demand a company's product offerings to "perform and transform," the governors, C-suite, and CEO must take risks to support PLSM. The interesting yet problematic issue for large, incumbent companies is that their shareholders are primarily traditional, old-school investment funds that demand high cash flow. In contrast, startups have shareholders who focus heavily on change and disruption.

- Seek to align the behaviors, attitudes, and beliefs of Governance team members with managing complex multi-product systems. Then, encourage them to guide product line teams using transparent feedback.

- Ensure the Governance team members are substantially involved, not just committed. I recommend securing more than 8 hours each month from each governor. This includes one 4-hour governance team meeting and four 1-hour "sponsorship" meetings with individual product line team members monthly.

- PLSM Governance is not for faint-hearted executives. These individuals must put aside their pride if the transparency being encouraged is to work as intended. Every member of the governance team will make mistakes at some point.

Governance and the New Efficiency Change Dilemma

A governance team must create opportunities for product line teams to improve their line's efficiency and Change Capacity. It's not one or the other; they must open pathways for both.

The governance role is crucial for running the Accelerating Change Race. Without effective governance, system inducement is likely to revert to pursuing only short-term efficiency gains.

Chapter 23 Summary
Enabling and Clearing Paths

- **Leadership Responsibility:** Governance is a leadership responsibility that enables product line teams to improve their line's Outcome Velocity.

- **Empowering, not Controlling:** Governance should empower and enable product line systems management, not limit and control them.

- **Ensure Flow Alignment:** An effective governance body should comprise senior managers who can ensure alignment, flow resonance, and ecosystem synergies.

- **Boundaries, Not Limits:** Governance teams should set enabling constraints and rational boundaries for product lines but avoid converting them into controlling limits.

- **PLSM Understanding:** A well-functioning governance team should understand product line systems management and clearly understand the business unit's strategy.

SECTION FIVE

The PLSM Framework

Components & Concepts | Doctrine & Principles

"You have to learn the rules of the game and then play better than anyone else."
- Unknown

Large companies must engage in PLSM. It is not a game to be avoided.

- The Author

Chapter 24

PLSM Components and Concepts

What Makes PLSM Tick

PLSM is a framework, not a process or method. It consists of components, concepts, a guiding doctrine, and several key principles that work together. I have discussed all of these in previous chapters.

Before discussing the doctrine and principles that guide the PLSM framework in practice, it's helpful to recognize the components and concepts involved in this management approach. Here, I present these fundamental elements in one list. In the next chapter, I will share the framework's guiding doctrine and the principles that bring the elements to life.

Components & Concepts

In Table 24.1, the left column lists framework concepts or system components, while the middle column provides a brief description of each. The right column indicates the primary chapter that explores and discusses each.

Table 24.1 PLSM Components and Concepts

#	Key Concept / Component	Description	Primary Book Location
1	**The Accelerating Change Race**	There is a continuing need to enhance offerings for efficiency and facilitate change to address new and future demands. The title reflects the intense, accelerating change that companies must confront. It serves as a deliberate and purposeful call to integrate significant change into the business's core product offerings.	Chapter 1 and throughout the book
2	**Context/System Substrate**	We need to shift our thinking from a single-product, orderly system context to a multi-product, complex system context. Our focus should be on product lines—specific segments of the market, technologies, key parts, enablers, constructors, ecosystems, constraints, and forces.	Chapters 2, 5-8
3	**System Flows and Emergence**	Every product line system contains within it many Information, work, and decision flows that are coupled to strategic moves and the emergence of outcomes.	Chapter 7; Online Appendix
4	**System Archetype**	The archetype is a category or group description of the system's architecture type. It dictates the system's dominant behaviors. The line's performance and Change Capacity depend on the system's behaviors and how well they align with a business model and the product line strategy. The system archetype stems from the combination	Chapters 9,10,11,12; Online Appendix

#	Key Concept / Component	Description	Primary Book Location
		of the product line's platform-levers and their type.	
5	**Performance Outcome Velocity Vector**	The Outcome Velocity vector is a three-dimensional visualization of a product line's performance, encompassing customer satisfaction, cash flow, competitiveness, direction, and movement.	Chapters 3, 4; Appendix 2
6	**Concurrent Verb & Noun-based Segmentation**	Verb-based segments are Jobs-to-be-Done customer need clusters (outcome clusters). Noun-based segments are traditional groupings such as geographies, industries, and demographics.	Chapter 9; Appendix 7
7	**Platform-levers and their types**	A platform-lever is a critical component (key part) of a product line system. A platform-lever contributes to the forming and delivery of multiple products within a line and by doing so provide each product with superior attribute "bang-for-the-buck." Platform-levers also enable teams to create individual products faster.	Chapters 9,11,12; Appendix 6,7
8	**Innovation Charters as enabling constraints**	An innovation charter serves as a target for a strategic move. It is utilized to assess the system's sensitivity regarding the move and to formulate a project for investment consideration. It also acts as a constraint that helps teams concentrate their efforts and choices.	Chapter 9,15; Appendix 7

#	Key Concept / Component	Description	Primary Book Location
9	**Centers of Technical Excellence**	A center of Technical Excellence is an organizational entity that forms insights related to a product line and business associated with a technology.	Chapter 20
10	**Product Line Strategy as an Enabling Constraint**	A product line strategy is an enabling constraint that offers managers and teams the guidance and focus necessary to drive growth. It clarifies "where to play' and "how to play."	Chapter 15
11	**Stack, Roadmaps, System Assemblage, and Assemblage map**	A product line stack is a static visualization of the system's key parts and how they relate to one another. The roadmap is a visualization of how the stack will change. The system assemblage adds non-key part agents such as enablers, ecosystems, and actors to the key part stack. The assemblage map shows all agent changes over time.	Chapters 9,10,11,12,22
12	**System Constraints and Forces**	Constraints limit flows (work, decisions, information, and outcomes. Forces can influence these flows both positively and negatively. Understanding, mapping, exploiting, and purposefully addressing these system flow influencers can lead to more impactful outcomes.	Chapter 14; Online Appendix
13	**Strategic Moves**	Strategic moves refer to actions that result in advancements, pivots, and transformations of a product line. All teams should aim to provide a cohesive	Chapter 16, Online Appendix

#	Key Concept / Component	Description	Primary Book Location
		series or stream of strategic moves that build upon one another.	
14	System Enablers	System enablers are agents that aid the system's flows. They include tools, practices, processes, and IT support.	Chapter 21
15	Constructors	A constructor is an enabler that creates or constructs a system's part or enabling agents. They operate independently of the system's day-to-day performance.	Chapter 21
16	Change Capacity and Composability	These are behavioral attributes of the system that indicate the ease of system change across all parts and agents.	Chapters 9,10,11,12; Appendix 5
17	Ecosystems	These are system enablers that function separately from the product line. Functional chain-links operate with the product line's parent organization. External ecosystems, such as supply chains and selling partnerships, operate independently of the parent organization. All product lines have such ecosystems.	Chapter 13; Online Appendix
18	Insight Generation and Sensemaking Affordance	This "affordance" is an environment that includes tools and aids that invite managers to work independently to discover impactful insights that have the potential to drive strategic moves.	Chapters 17,18

#	Key Concept / Component	Description	Primary Book Location
19	**Distributed Initiative and Leadership**	This leadership style is self-directed and motivated by personal initiative, focusing on advancing the entire product line system or its parts and agents.	Chapter 19
20	**Product Line System Governance**	Governance is the oversight, guidance, and pathway clearing for product line systems management.	Chapter 23; Online Appendix
21	**Product Line Congruence Theory**	This theory posits that for a product line and a business to operate efficiently, the product line archetype and the business model type must align.	Chapter 11, Online Appendix
22	**Archetype Disruption Theory**	This theory posits that industry disruptions are possible only when the innovation comes from a new-to-an-industry system archetype, regardless of the originating company's size or culture. Other conditions are also necessary.	Chapter 12; Online Appendix
23	**Product Line Change Theory**	This theory posits that product line systems change more rapidly and easily when proper changes are made to system agents (enablers and ecosystems) at the system's lowest granular level. Such changes should accompany the changes that teams make to the system's key parts and stack.	Chapter 22; Appendix 5

The Full System, Not The Sum of Its Parts

It's the entire system that matters. While we want each part and agent of a product line system to perform well, the interaction of all parts—the totality of the system—affects the line's performance and capacity to change. While we must recognize the presence or absence of each part or agent, we mustn't fall into the trap of thinking that the system's behavior is merely the sum of its parts' behaviors.

It's common to hear managers adopt a "reductionist" line of thought when explaining a product line system, detailing the individual components that make up the line.[84] However, product lines are complex systems and are not reductive. We can't break the system down into its parts and agents and expect that examining each one will tell us how the system will behave.

While we strive to understand each part of the product line and its potential impact on the system, the overall behavior of a product line does not easily come together. We need to observe the system's behavior for a clearer understanding. Yes, we aspire to comprehend every component of the product line, but we shouldn't assume that this knowledge translates directly to the system's behavior.

Three Important Theories

In this book, I proffer three theories. Individually and together, they have significant implications for product innovation and management. Each theory emerges from the repositioning of product innovation and management within the context of multi-product complex systems. I provide a high-level overview of the theories here to emphasize the solid foundation and the richness of this context shift.

1. Product Line Congruency Theory

A product line's archetype and an organization's business model must operate in congruence for the product line system and business to operate efficiently. Consequently, any change to one requires a corresponding change in the other.

Ongoing operations require teams to ensure their product line behavior aligns with the business unit's behavior. The product line systems archetype is a critical determinant of the line's behavior, and the business model guides the functional interplay and behavior of the organization. Therefore, to attain superior performance, the product line system archetype must align with the business model type.

See more in Chapter 11 and the online appendix.

----

2. Archetype Disruption Theory:

Industry disruptions occur only when innovation originates from a system archetype that is new to the industry, regardless of the company's size. Also, for new system archetypes to succeed and expand, they must align with the corresponding business models and strategies.

This theory suggests that a company's innovation can disrupt an industry when it introduces a product line system archetype that is new to that industry. Notice the importance of viewing disruption from a complex multi-product systems perspective, rather than the traditional single-product orderly systems perspective.

See more in Chapter 12 and the online appendix.

----

3. Product Line Change Theory:

Product line systems change more rapidly and efficiently when requisite changes to enablers and ecosystems, at their finest level of granularity, accompany key part changes.

This theory suggests that you need to modify the system's structure to achieve the greatest performance gains from changes to the product line stack, including scaling any part of the system. Furthermore, this change should focus on the system's lowest level of granularity: the system's agents, enablers, constructors, and supporting ecosystems.

See more in Chapter 22 and Appendix 5.

As you step into PLSM, be patient and stay open-minded. It may take time for individuals to adapt to thinking about product lines as complex, multi-product systems.

PLSM Concepts and Components and The Accelerating Change Race

Deliberately applying PLSM concepts and components enables a company to remain competitive in the Accelerating Change Race.

Chapter 25

PLSM Doctrine & Principles

Direction, Capacity, and Change

P LSM is not a standardized process, method, or template. Instead, it is a collection of components and concepts put into action, guided by a doctrine and a set of principles. It creates a complex, multi-product system context for product innovation and management, transforming how people think about and engage in innovation and product management.

Context is Everything

In this chapter, I outline the PLSM guiding doctrine and principles to achieve impactful gains. They are based on sound theory and informed by practical experience and case studies. While the principles may appear distinct, they overlap and complement each other. It's the complete set that matters, not just one or two.

I do not intend for the doctrine or principle to be unchangeable or final. Instead, I present them as foundational steps toward a dynamic and effective approach to product line systems management and addressing the NEW efficiency change dilemma.

Remember:

PLSM is not copyable. While the doctrine remains firm, you must adapt its principles to your specific situation. No two product lines or circumstances are the same. Each organization's approach to PLSM will be unique to its product line and company. The nuances and importance of each principle will evolve as the product line system changes.

The Guiding Doctrine

Three tenets working together create the PLSM doctrine, forming the foundation on which the framework's principles stand.

1. **The Core Must Change. The Accelerating Change Race Demands Increasingly Greater Change to Core Offerings.**

The continuous acceleration of change driven by Artificial Intelligence requires companies to undertake significant transformations in their core products. This demands intentional responsiveness to the forces and constraints related to technological advancements, emerging customer needs, competitive dynamics, and regulations.

The Accelerating Change Race underscores the need to redefine product innovation and management strategies to transform core offerings effectively.

2. **Product Line Systems are The Most Important Entity In A Business. We Must Focus on Them Accordingly.**

Focus on changes and advancements in the product line system rather than the performance of individual products or business units.

Focusing on product lines acknowledges that effective strategies are interconnected. Product lines provide the ideal foundation for implementing change and delivering a consistent series of impactful strategic moves. A strong emphasis on a product line enables a business to enhance customer satisfaction, cash flow, and competitive advantage—the line's Outcome Velocity—both in the short term and the long term.

3. Product Lines Are Complex Multi-Product Systems. We Must Manage Them As Such.

Product lines should be managed as complex systems. Teams need to recognize the interconnectedness of their components, agents, constraints, and the forces at play. This focus enables product line teams to navigate uncertainties and leverage cross-product synergies.

A shift from single-product to multi-product thinking enables companies to maximize leverage, responsiveness, scale advantages, and complementary product differentiation. The transition from orderly systems to complex systems thinking involves moving from an emphasis on root cause analysis and management by objectives to fostering flow enablement and the emergence of outcomes.

Avoid discarding single-product or orderly systems thinking in favor of complex multi-product thinking. Understand when and how to apply each approach to optimize strategic actions and local subsystems.

Key Principles

UNDERSTAND AND MANAGE THE SYSTEM'S BEHAVIOR AND CHANGE CAPACITY

#	Principle
1	Understanding platform-levers is crucial. They define the system's archetype and behavior. Deconstruct and reconstruct your conceptual model of the product line system's structure and architecture with a focus on its-existing and potential platform-levers. Investigate the system archetypes (new and prospective) and the resulting system behaviors. Do this regularly.
2	All changes in the product line system, whether incremental or radical, scaling up or downsizing, demand system agent changes at the most granular level. (Product Line Change Theory)
3	Develop a process to support your product line team in creating a compelling Change Capacity narrative. Conceptually deconstruct your line to investigate and assess its resistance to change at the system's most basic level of granularity. Then, rebuild the system model in ways that enhance the system's Change Capacity. Learn from this effort before altering the real system. Constraints and forces define the system and influence its ability to change key components and supporting agents.
4	Build and continually increase Change Capacity before you need it. Change Capacity is inversely related to the cost of change. The lower the cost of change, the higher the Change Capacity.
5	A product line's archetype, strategy, and business model must work together. If one changes, the other two must also change—conflicting archetypes can jeopardize businesses. Recognize when to separate product line systems.
6	A strategy should concentrate on the 10% or less of the system agents and forces that drive 80% or more of a product line's outcomes. To enhance the 10%, maintain an intense focus on it. Achieve this through enablers, organizational structure, and roles and responsibilities.
7	Establish and communicate the product line system's primary "enabling constraint"—the line's strategy—and then promote its progress and adaptability to industry, market, technological, and organizational changes.
8	Gain advantages from Internal (Chain-link) and External Ecosystems.

FOCUS ON IMPROVING THE LINE'S OUTCOME VELOCITY THROUGH STRATEGIC MOVES THAT DELIVER COMPOUNDING GAINS

#	Principle
9	Play with a full system (of parts and agents) and more than one product. Recognize that changes to the system agents, particularly enablers and constructors, are as important as changes to the product offerings.
10	Shun the idea that you need only to speed up product development or increase the pace of incremental innovation to be successful. That is no longer true.
11	Focus intensely on the interplay between your product line system's most important parts, agents, and the system constraints and forces. Create this focus on the system's parts and agents at their lowest level of granularity.
12	Target market segments better than competitors can engineer attributes to position products before messaging the positioning. But don't rule out already-served or underserved needs. Both can be targets for beachhead products and ramping up a line's Outcome Velocity.
13	Analyze each strategic move before and after its rollout to determine if there are local orderly subsystems that might or are hindering outcome emergence. Aggressively address potential and actual hindrances using orderly systems methods such as the Theory of Constraints. Exploit, don't throw out, orderly system methods. They can be invaluable for ensuring successful strategic moves.
14	Evaluate potential outcomes and strategic moves based on the energy and time teams need to carry them out. Compare this to the gain they believe may come with its success.
15	Add and continually improve enablers to your product line system, but only after redressing current practices to match PLSM.
16	Exploit artificial intelligence in favor of the product line system, not in favor of single-product innovation and management.
17	Never start a project directly. As a form of due diligence, begin with charters that articulate boundaries and expansion that best fit the product line roadmap (Outcome Velocity impact and strategic move coherence). Form the project based on the desired charter.
18	Exploit the system by gaining and nurturing coherency across all system enablers. Create and employ enablers and constructors that amplify the impact of current and potential system agents, parts, and flow.

TRANSITION INTO AND EMBRACE PLSM

#	Principle
19	Organizations must transition into PLSM. You can't "implement" it. PLSM has no on-off switch.
20	We must integrate flow principles with systems thinking. Workflows, information, decisions, and outcome flows intersect within the product line system, creating a dynamic interplay. As a general rule (mostly, but not always), improved flows tend to enhance the system's performance.
21	Seek flow resonance where one flow enhances another flow beyond coherence and alignment. Avoid, remove, or mitigate constraints that impact all flows—workflows, information flows, and the strategic move outcome flow. Aim to leverage flow resonance.
22	Multi-product systems management must connect to organizational agility. The context shift begins with leadership and encompasses all managers and contributors.
23	Insights represent the most crucial information. Cultivate insight and sensemaking by leveraging affordance, distributed leadership, and organizational structure; new knowledge and insights form the lifeblood of the system. The most significant insights emerge at the intersection of technologies and customer needs. Strive to eliminate all barriers to insight generation and sensemaking.
24	Decisions, judgments, and choices should rely on insights and principles rather than consensus. Use the product line strategy as a boundary-defining enabling constraint for the decision flow.
25	Highlight the visual communication aspects of your system, including its components and dynamics. Interactive maps are superior to interactive text documents.
26	As a general rule (mostly, but not always), exploring how the system might be influenced toward a novel assembly today is more valuable than future exploration.
27	Understand your heuristics and assumptions. Clearly outline them. They assist managers in making decisions without having to relearn.
28	Hubris kills; avoid the hubris success trap.

Discussion Points: Unifying the Principles

DP1: Managing multi-product systems must connect with organizational agility. The shift in context begins with leadership and encompasses all managers and contributors.

Product line and business unit leaders must embrace and guide the entire product line system, not just its components. PLSM becomes established within a business only when business unit leadership reviews product line roadmaps, force field maps, and the rationale and coherence of how the interplay of system parts and potential strategic moves will drive a line's Outcome Velocity improvement.

Leaders should promote insight generation and enable sensemaking to enhance the collective understanding of interactions among current and potential system components and agents.

Do not confuse product line governance with decision-making. Governance must adopt and adapt a strategy and establish guidelines to improve Outcome Velocity. Decision-makers must manage judgments, choices, and decisions across all components and workflows of the system.

Revise your tools and practices (including templates and processes) to align with multi-product system thinking and the potential incorporation of new components and forces. Seek change that improves your team's capacity and skills to carry out:

- Weak-signal Sensing

- Sensemaking and insight generating

- Safe-to-fail Experiments

- Strategic Move Creation and Coherency

- Understanding Current and Potential Key Parts and Agents

DP2: Focus intensely on improving your product line's Outcome Velocity.

Promote a careful and strong focus on the Outcome Velocity of your product line. Do not allow time-to-market KPIs and project management throughput measures to overshadow the importance of your product line's Outcome Velocity.

Focus on enhancing the overall system's Outcome Velocity rather than optimizing individual projects. Keep striving for improvements in both near-term and long-term product line Outcome Velocity. Emphasize both efficiency and change.

Insights related to actions will drive improvements in Outcome Velocity. All contributors must engage in forming, compiling, and communicating these insights. Product line teams must take responsibility for exploring and utilizing all insights through scenario analyses and strategic move creation for both near- and long-term enhancements of Outcome Velocity.

Leaders and managers must take responsibility for eliminating or minimizing all obstacles to gaining insights and determining the actions or strategic moves that most effectively drive Outcome Velocity. They must leverage organizational agility—through temporary structures, practices, and system supports—to cultivate insights across system components and develop coherent strategic moves.

DP3: Deconstruct your product line assemblage often; Explore new assemblages and architectures using system assemblage mapping.

Understand the core of your product line system: its key parts, forces, actors, constraints, enablers, and constructors. Concentrate on the level of consensus surrounding the system's platform-levers and archetype. The more effectively you perform this work, the greater your improvement in Outcome Velocity will be.

Deconstructing your product line system is the greatest source of learning and insight for your team. Make the most of this.

DP4: Avoid, remove, or mitigate constraints that impinge on all flows—workflows, information flows, and the strategic move outcome flow.

Create urgency for system advancements and Change Capacity before an emergency demands it. Utilize orderly system constraint mitigation methods, such as root cause analysis and the Theory of Constraints, to support distinct sections of the product line's flows. Anticipate flow impediments that affect strategic moves before executing the move.

Assess flow constraints pertaining to each system component and agent. Methodically eliminate or lessen these constraints to improve the emergence of outcomes and increase Outcome Velocity.

DP5: Add and continually improve enablers to your product line system only after redressing current practices to match PLSM.

- Do not let single-product, orderly tools and practices (the enablers) control the product line system and its flows. Ensure that the use of orderly system tools and practices remains within the PLSM framework and is managed by the product line team.

- The most common weakness of product lines is the failure to possess a complete and effective stack. The components most often missing are Verb-based segments and platform-levers. A product line stack's effectiveness is frequently undermined by poor quality of key parts or the lack of coherence among the key parts.

- Establish measures for product line Outcome Velocity and develop dispositional futurecasting models. Analyze the three dimensions of Outcome Velocity: cash flow, customer satisfaction, and competitiveness. Consider employing iterative Delphi techniques to formulate informed best-guess futurecasts.

- Product line roadmaps should convey insights about key system parts. These parts include Noun and Verb-based (Jobs-to-be-Done) market segments and platform-levers. They also include products currently in the market, products in development, and product innovation charters. Plus, roadmaps should present technology building blocks, possibly with details about their maturity and readiness.

- Teams should perform frequent risk and uncertainty assessments across the entire product line system. They should also conduct these assessments for groups or sets of parts and forces within the system. (When many development projects share the same risk, the risk is systemic. It's very difficult to address systemic risk at a project level).

DP6: Highlight the visual communication of your system, including its components and dynamics. Interactive maps are superior to, but should not negate, textual documents.

Visually represent the current and future state scenarios of the system's components and forces using product line roadmaps and force field diagrams. Encourage two-way communication about the diagrams with project team members and product line leaders.

Encourage communication among functional groups and other product lines. Foster communication throughout the organization's hierarchy. Actively engage with key suppliers and partners, expecting their enhanced contributions.

DP7: Establish and communicate the product line system's primary "enabling constraint"—the line's strategy—then promote its development and adaptation.

A product line strategy is the main enabling constraint of the system. It guides and influences the system's flows and the emergence of outcomes, enhancing the velocity of product line outcomes.

Create this enabling constraint through an iterative process with essential system actors and contributors. All information and insights should be easily accessible to all contributors.

Ensure the governance body commits to a strategy and its ongoing advancement and evolution. Clearly communicate the rationale for the strategy and any changes to all contributors.

Establish a strong focus within the boundaries set by the system's enabling constraints. Achieve this through clearly defined organizational structures, roles, and responsibilities. Refine and emphasize the strategy within cross-functional processes and practices. Align the strategy with improvements in Outcome Velocity while acknowledging the line's challenge of efficiency change. Stay flexible and adjust as needed to maintain

357

enhancements in Outcome Velocity. Enabling constraints should be intentionally adaptable rather than rigid and unyielding.

DP8: Exploit the system by fostering and maintaining coherence among all enablers.

The greatest benefits arise from identifying the best product line system to yield the best results, enhance speed, and make the most strategic decisions to implement this. Such planning elevates concepts and designs above products. Charters (targets) for innovation and change become more crucial than concepts and designs.

Graphically lay out and constantly update a coherent set of strategic moves, each defined by well-thought-out innovation and change charters. The coherence of these system parts is the primary driver of outcomes and Outcome Velocity.

DP9: Develop, strategize, and execute an ongoing series of coherent and cumulative actions.

Employ creative and strategic thinking to develop potential strategic moves. Incorporate insights and lessons as material for generating new moves. Next, evaluate the velocity and impact of these potential moves. The objective is to identify the most effective approach to address the issues and improve Outcome Velocity results.

- Create scenarios to explore strategic moves and the likelihood of outcomes. Use insights from scenario explorations, forecasting, and back-casting to generate insights that can inform and refine strategic moves and their overall impact

- Develop and continually enhance a discipline to pursue each platform-lever's next-generation advancements.

- Build out and test platform-lever replacements before they are needed.

DP10: Never start a project directly.

Never start a development project directly. First, convert the concept into an innovation charter. Then, conduct due diligence regarding the charter's impact on the system and its Outcome Velocity. Next, revise the charter and use it as a target for innovation or concept creation. Ensure that the charter belongs on the product line roadmap.

DP11: Seek flow resonance through flow coherence and alignment.

Aim for resonance across workflows, decision flows, and information flows. The goal should be to boost the flows and improve the emergence of outcomes. Compile and test key assumptions and heuristics to verify their effectiveness. Then, adjust, remove, or refine the assumptions and heuristics to achieve optimal flow alignment resonance.

Flow resonance is a state where each flow is progressing better than it could if agent changes were made to that flow alone.

DP12: Gain Chain-link and Ecosystem advantages.

Achieve alignment across organizational chain-links to speed up Outcome Velocity. Drive changes within functional groups that will enhance your line's Outcome Velocity. Review and consider modifications to your line that allow other chain-links to amplify their contribution to your line's outcome emergence.

Actively pursue ecosystem contributions that can enhance the line's Outcome Velocity in both the short and long term. Ensure ecosystem alignment and boost Outcome Velocity. Engage current and potential ecosystem players in developing and executing strategic initiatives.

DP13: Concentrate intensely on the interaction of the key components and agents of your product line system concerning system constraints and forces.

Recognize that what you considered important in the past months may not hold the same significance now. Adjust your focus as necessary. This work demands clarity and agility. Establish processes, structures, and tools to support both.

DP14: Form and employ enablers and constructors that amplify the impact of current and potential system agents, parts, and flows.

The product line team and all contributors should consistently work to improve the speed of product line outcomes. This is accomplished not only through product offerings but also by utilizing enablers and constructors.

Teams should strategize, design, and develop enablers to improve the line's Outcome Velocity. They should also focus on creating constructors that build system components and execute system flows. Constructors may be the most critical element of your product line system.

DP15: Target market segments with greater specificity and proficiency than competitors.

Seek an advantage over competitors by leveraging your skills to create and target Noun and Verb-based market segments. Build and continuously improve tools that support work, generate insights, and aid decision-making related to this skill.

Though aligning product attributes with customer-desired outcomes (Jobs-to-be-Done) is essential, developing and redefining market segments more effectively than competitors can provide a lasting strategic advantage.

PLSM Principles and The Accelerating Change Race

PLSM principles create a framework that intentionally guides organizations in their work, decisions, and orientation, helping them remain competitive in the Accelerating Change Race. They assist core product lines in adapting and thriving amidst greater change.

The principles are based on complex systems theory, knowledge management, and multi-product thinking. They are further developed through product innovation and management practices. Marketing, organizational dynamics, technology, strategy, and knowledge of innovation refine and shape the principles.

I offer the principles as a solid foundation for PLSM, not a complete and final set.

Chapter 25 Summary

The PLSM Framework & its Principles

- **The PLSM Framework**: Product line systems management is a framework of principles, not a process or practice. This chapter offers twenty-one principles that work together and are not isolated from one another.

- **Adapt, Don't Copy:** The principles are pliable, not fixed. They must be adapted to fit the organization and its product lines and adjusted as the organization and product line situation changes.

- **Guide, Don't Control:** Decisions within complex systems should be guided by principles, not based on rigid criteria. In product line systems management, the principles seek to guide, not control, a product line's outcome emergence and Outcome Velocity improvement.

- **Evolving Framework:** The principles laid out in PLSM are not intended to remain static. They must evolve as we learn more about all product lines' multi-product, complex nature. It is OK and quite healthy to explore other principles and add to and subtract from the principles offered in this chapter.

Chapter 26

Transitioning Into PLSM

Recasting Product Innovation and Management into Product Line Systems Management

This chapter provides a broad overview of transitioning to PLSM for an existing product line within a large organization. However, each transition must be tailored to fit the specific context. My consulting firm provides training, guidance, facilitation, and one-on-one coaching in PLSM. You can also find valuable tools and templates at the online appendix

PLSM is not a process like Phase-Gate development or a method like Agile. Instead, PLSM comprises principles and approaches combined with a contextual shift. Together, these form a framework. Organizations must transition into the PLSM framework; you don't "implement" it. PLSM does not have an on-off switch.

The transition from traditional product management to product line systems management requires an openness to learning and a commitment to new principles, methods, and orientation. This shift must involve individual managers and executives, as well as entire departments and teams. The key is shifting the context in which everyone thinks.

The PLSM context emphasizes a multi-product approach rather than a single-product approach. It encourages managers to view product lines as complex systems rather than orderly ones. It also instructs each product line team to manage multiple flows and to expand their focus beyond just project workflows.

Starting Up

At the start of transitions, terms like "platform-levers," "verb-based market segments," and "innovation charters" may be unfamiliar to your organization. This indicates that there is no existing process or support system in place for these components to thrive and add value to your product line. Therefore, you'll need to adjust your current processes and practices to fully benefit from these multi-product elements.

Please recognize that the transition must address several organizational challenges, and a hierarchical structure may limit your line's ability to adapt. The new management context might also require adjustments to specific roles and responsibilities. Moreover, in PLSM, generating insights should be nurtured by promoting independent initiatives and distributed leadership. While it may be easy to list these changes, implementing them can be difficult. There will always be resistance and objections to changing practices and approaches in large organizations.

For many product lines, transitioning to PLSM necessitates significant changes. The challenge lies in the fact that altering everything at once can negatively impact the product line, the company, and numerous careers. Moreover, the transition may stumble and even fail if changes are implemented too slowly.

Top Management Involvement: The Oversight Team

A key lesson learned from major cross-product development initiatives, such as portfolio management and product line roadmapping—both important enablers in PLSM—is that the work cannot be orchestrated solely from a bottom-up perspective.[85] Gaining top management involvement at the outset of the initiative is crucial. The initiative is unlikely to succeed if the business unit's leadership team does not support the effort through their participation and meaningful funding.

A top management oversight team must support and guide the transition. During major changes, such as the transition to PLSM, senior management's commitment can be short-lived. Instead, transitions should aim to involve top management.

An oversight team must have strong leadership and coaching skills to effectively guide those implementing the transition. Ideally, the oversight team should include two C-level executives: one from marketing and the other from technology. This duo may also be responsible for leading the PLSM governance team.

The Transition Team

The transition team, which will handle most of the transition work, should consist of those managing the current products and product lines. Team members may also include individuals involved in supporting product innovation and management, as well as managers from relevant chain-link functions.

Their job is to teach, guide, and influence PLSM's acceptance and ongoing management. This requires getting deeply involved in all systems agents with all system actors. While outside experts may assist with certain aspects of the initiative, the team should be responsible for completing the transition and preparing product line teams to take charge of improving the velocity of their line's outcomes.

Transition Phases

I see the complete transition as consisting of several phases, each with distinct orientations and purposes. See Table 26.1. Each phase overlaps its beginning with the conclusion of the preceding phase, intentionally building on earlier accomplishments.

Teams address the components of the product line system, including key parts, enablers, flows, and the necessary organizational structure, roles, and responsibilities, as well as chain-link alignment during each phase. A table in the online appendix outlines the work to be completed on each component during every phase. This table serves as a starting point for teams to modify as needed to better accommodate specific circumstances.

Table 26.1 Transition Phases

Phase	Purpose	Description
Starting The Transition	Build an Understanding and Awareness of PLSM	Establishing an understanding of multi-product complex systems and the benefits of PLSM
Baselining Current State Assemblage & Performance	Anchor a View of the Current Situation and the Line's Assemblage	Analyzing the product line system, identifying its constraints, enablers, ecosystems, and key parts
Future State Envisioning	Explore Possibilities and Embracing Potentials	Creating a vision of a future state through narratives and stories, coupled with trends and discontinuities

Phase	Purpose	Description
Confirming the System and Exploiting PLSM	Set Up System Change Navigation; Build Change Capacity into the System	Enhancing system agents and key parts, focusing on strategic moves to advance the system to drive Outcome Velocity and improve Change Capacity
Maximizing the Product Line's Outcome Velocity	Manage the Advancement of the Ongoing System	Analyzing and considering significant changes to the system architecture through system archetype moves
Amping up the System	Radically Improve Outcome Velocity via Constructors and Platform-levers. Exploit PLSM to the fullest	Creating, revamping, and investing in enablers, constructors, and platform-levers.

Starting the Transition

The best way to start a PLSM transition is with an explosion. Let me explain.

In college, my engineering curriculum required me to take a course on gas turbines. Each turbine has blades designed to capture and be pushed by the mass released during gas combustion. When you know the force of the emission, as well as the size and shape of the blade, a novice engineer can calculate the turbine's rotational speed. This seemed to me to be what I'd call a chaotic (not complex) system.

The students' problem was not understanding the turbine's ongoing spin. Our issue was that no one knew how to calculate the forces required to start the turbine without blowing it up in an explosion. To me, a gas explosion is chaotic, not orderly, and I'm unsure how to calculate the forces from such an explosion. I've since learned that experienced turbine engineers perceive the explosion as more complex than merely chaotic.

Like the turbine, transitions into PLSM require an initial explosion to get things moving. Consider two recent explosions that affected the dynamics of change for product lines. Elon Musk fired nearly the entire 500-person Tesla Supercharger product line team in a dispute over eliminating jobs to prepare for a major archetype shift towards autonomous driving. Meanwhile, Google's core search business was stunned by the launch of OpenAI-Microsoft's ChatGPT, while its LLM AI continued to languish as an H3 technology (as per McKinsey's Technology Time Horizon model[18]) in Google Labs. One explosion was driven by leadership; the other was prompted by competition. Both resulted in significant change.

Concerns at the Start

PLSM Startup teams must address several concerns. First, key players need to recognize that their current approach to tackling the combined challenges of efficiency and change is ineffective and, according to the equilibrium equation, is painful. Second, they must view PLSM's contributions as positive, which will lead to a gain in Outcome Velocity. If these factors are not in place, the transition will face difficulties. This makes teaching product line Outcome Velocity versus traditional time-to-market KPIs a crucial first step.

The biggest roadblock in the transition is a lack of awareness and understanding about PLSM. When managers and leaders fail to grasp the basics of the topic, it becomes challenging to explore effective strategies and approaches with them. The PLSM lexicon is unique, and managers often misconstrue the topic when they encounter a word or phrase without understanding the context of PLSM. This disconnect can undermine the initiative.

Leadership teams may not recognize that single-product orderly systems thinking is a significant or costly problem. Instead, they perceive the old paradigm as normal and feel no urgency to change. This blind acceptance of the old approach must be addressed before a company can fully benefit from PLSM.

Identifying and understanding a product line's platform-lever(s) can be challenging for some companies, particularly those with established product offerings. This challenge arises from the lack of a platform-lever or its transformation to the point where it provides minimal leverage. The solution to this situation is to navigate through phases of transition with a "best guess" platform-lever and allow PLSM cycles to reform or target the creation of platform-levers while adjusting for changes in the system's behavior (see Chapter 11).

In all cases, awareness and understanding provide the foundation for strong PLSM.

Continuing the Progress

While an explosion can be effective in initiating the transition, more is required once the dust, if any, settles.

Those experienced in company changes and transformations may recognize essential behavioral influences on the initiative. People change, and the transition can only continue when each individual, group, or department perceives that the benefits of the new approach outweigh the drawbacks of the old one. The economic and psychological costs of the change must be outweighed. This is change management 101, as laid out by Kurt

Lewin many decades ago. He summarized it in a Change Equilibrium Equation. See Figure 26.1[86]

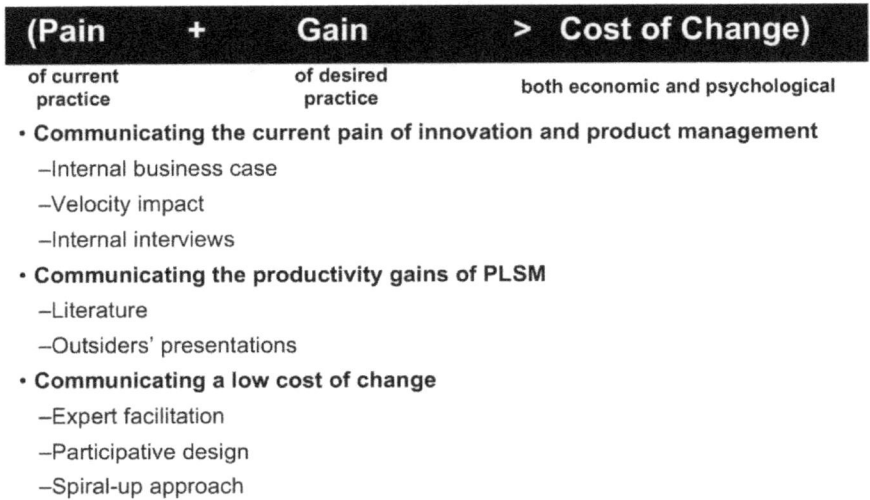

(Pain + Gain > Cost of Change)

of current practice	of desired practice	both economic and psychological

- **Communicating the current pain of innovation and product management**
  - –Internal business case
  - –Velocity impact
  - –Internal interviews
- **Communicating the productivity gains of PLSM**
  - –Literature
  - –Outsiders' presentations
- **Communicating a low cost of change**
  - –Expert facilitation
  - –Participative design
  - –Spiral-up approach

Figure 26.1. The Change Equilibrium. *For an organization to move forward with PLSM, individuals and groups must recognize the pain of the current situation. They must then weigh that against the benefits they perceive from PLSM. Change will happen when they see this balance as outweighing the psychological and economic costs of the change.*

During each phase of the PLSM transition, the primary goal is to tilt the equilibrium in favor of change. This advantageous tilt is vital not only for each product line team but also for all other contributors to the product line, including the company's functional groups (chain-links), as well as external partners and the supply chain.

Most importantly, the equilibrium must tilt favorably for all system actors and contributors during all phases of the transition. If the tilt becomes unfavorable, the transition is likely to falter or come to a standstill, potentially causing the organization to revert to outdated single-product innovation and management.

Transition Durations

The time an organization takes to transition to PLSM can vary significantly based on the complexity of the job. To estimate this, I first assess the number of Noun-based and Verb-based segments, which I refer to as the "width" of the product line. The broader it is, the longer the transition will take. Second, I evaluate the number of SKUs in the line, known as the "length" of the product line. Larger product lines generally require more time to transition than smaller ones. The third dimension is the total investment made into platform-levers, referred to as the "weight" of the line. The greater the product line's weight, the longer the transition will take.

My rule of thumb is that very high weights, widths, and lengths can result in the transition taking over a year to complete. In contrast, lower values can significantly shorten transition times to less than one year. I offer these values to provide you with a rough estimate of the duration for initial planning, not to dictate the pace of the transition.

The deliberate induction of change in the context of product innovation and management must be accompanied by support and activities to facilitate it. These need to be specific to the product line and organization. While organizations should undertake as much of the transition work as possible, it is preferable to complement internal efforts with external expertise. I know, that's a plug for consultants, but reducing the time and effort required to implement the transition is well worth it.

Moving Forward

Every topic discussed in this book should be included in the transition. To assist you in transitioning to PSLM, I recommend visiting the online appendix for additional information.

Transitioning to Run in The Accelerating Change Race

Embracing PLSM is not only about implementing a process or establishing a practice; it requires shifting the context of work and decision-making from single-product, orderly systems thinking to multi-product, complex systems thinking.

This transition is essential for addressing the demand for both simultaneous efficiency and change while avoiding the negative innovation loop. Without PLSM, organizations will continue to face challenges as technology and market changes accelerate.

Chapter 26 Summary

Transitioning Into PLSM

- **PLSM Framework Overview:** Product Line Systems Management (PLSM) is not a specific process but a framework of principles designed to manage complex, multi-product systems. Transitioning into PLSM requires organizations to rethink how they manage their product lines, moving away from single-product, orderly systems.

- **Transitioning Challenges:** Shifting to PLSM involves overcoming significant challenges, such as changing organizational mindsets and roles. The transition demands active involvement from managers, executives, and teams to adapt to the new multi-product system context.

- **Top Management's Role:** Successful PLSM transitions require vigorous support and involvement from top management. Their participation is crucial in guiding the transition team and securing the resources and authority to implement changes.

- **Transition Phases:** We divide the transition into PLSM into several phases, each building on the previous one. These phases include understanding the current system, envisioning future possibilities, and setting up changes to improve product line Outcome Velocity.

- **Importance of Awareness:** One of the biggest obstacles to implementing PLSM is a lack of awareness and understanding among managers and leaders. Educating them on the benefits and requirements of PLSM is essential for the transition's success.

- **Change Equilibrium:** The transition must shift the organizational equilibrium towards change by making the benefits of PLSM outweigh the costs. This requires visible and impactful actions to gain momentum and prevent the organization from reverting to old practices.

- **Transition Duration:** The time needed to transition to PLSM fully varies depending on the complexity of the product line. Factors such as the number of products, the extent of required changes, and the investment in -levers influence the transition's length.

SECTION SIX

Corporate Innovation & PLSM

Radical innovation is essential, not optional, for large corporations. Its complex nature requires that companies rethink and reform their approaches to such innovation, rather than imitating that of others.

- The Author

376

Chapter 27

The Urgent Need to Improve Corporate Innovation's Effectiveness

Corporate Innovation (CI) is distinct from core product innovation. The previous chapters focused on core product lines. In this chapter and the next, I will explain the reasons and methods by which CI should leverage PLSM to improve its effectiveness.

C orporate innovation refers to a large company's strategy for significant innovations. Its goal is to create growth opportunities that are separate from core product offerings.

However, CI encounters a major issue. At present, CI is notably ineffective.

The Common Narrative

It's not that big companies can't create innovations outside their core business; they can. The issue is that most of these innovations—some argue as high as 97%—fall significantly short in scale and success.[16]

Grand stories have become the folklore that inspires managers to initiate and implement CI activities. These narratives build credibility by referencing well-known companies and discussing long-term societal benefits. They do not provide ROIs or equity value gains. Good intentions drive this approach. However, neglecting financial evaluations overlooks reality.

Corporate Innovation can create theater. Such staged venues are often referred to as Hubs and Labs.[87] CI leaders collaborating with their top management aim to shift the investment community's view that the company is outdated, lacking innovation, and undeserving of higher equity valuations. No management team wants to be associated with these characteristics.

Regrettably, CI efforts produce minimal meaningful success. Instead, these initiatives showcase innovative technologies and suggest that they will deliver value in the future. Too often, these promises fall short, causing top management to halt the work. In 2024 alone, major companies like Walmart, SAP, and AXA ceased their corporate innovation efforts.

Can versus Can't

Some in the CI community promote a narrative suggesting that a core business cannot change. They argue that those managing existing products excel only at driving efficiency, not at fostering innovation. However, core product lines can innovate when teams intentionally build their line's capacity for change. This message serves to justify CI's existence.

It's not that core product lines "can't" change. The issue is that many existing product lines have yet to undergo significant changes. And there is a significant difference between "can't" and "have yet to."

Leadership Understanding

The CI community recognizes that most initiatives fall well short of acceptable success rates. If these initiatives were genuinely successful, the community wouldn't need to create a false narrative, and senior management would openly endorse CI.

I have dedicated this book to explaining why and how core product lines must evolve, as well as how PLSM facilitates this transformation. The option to avoid changing core offerings has long since ceased to exist.

When a company's leadership accepts the misleading narrative about core product lines, it mistakenly encourages these lines to favor efficiency over advancements and changes. Given the low effectiveness of CI, this narrative exacerbates the negative innovation cycle mentioned in Chapter 1. This situation can harm the entire business.

Issues and Finger-Pointing

Some CI advocates communicate a similarly flawed message. They assert that top executives—particularly CEOs—don't understand CI and resist or reject it, leading to its underperformance. They argue that top leaders exacerbate CI's challenges by focusing on short-term results and showing intolerance for uncertainty and risk.

You'll also hear some CI experts argue that top executives often overlook the substantial value generated by new companies like

Tesla, Snapchat, and Roku. We're led to believe that CEOs fail to recognize that these valuations significantly exceed the value of the parent company's core offerings.

The challenge for the CI community is that CEOs have come to realize that CI methods and approaches do not generate significant value. Many CEOs acknowledge this situation and question the rationale behind investment decisions in corporate innovation.

Gaining universal approval from CEOs is only possible through success, and the current approaches to CI do not warrant such support.

Elusive Effectiveness

Practitioners, academics, and consultants have devoted decades to tackling the poor performance of CI. They have pursued three strategies to overcome the lackluster outcomes. The first two focus on how work is conducted, while the third seeks to increase the number of projects in progress.

The first strategy is to minimize CI's interaction with the core business. It deliberately separates CI decisions, investments, and activities from the business's primary revenue source. This approach is sensible if you believe a core business should limit innovation efforts to short-term incremental advancements. This ring-fencing strategy utilizes Lean Startup, Skunk Works, Innovation Hubs, and Business Development Labs to organize and carry out the necessary CI work.

These ring-fencers argue that blended ambidextrous organizations harm CI and the core business. Ambidextrous organizations are those where the core business and new-to-the-world entities operate side by side. They contend that ambidexterity should be viewed solely from a legal and financial perspective, rather than from a shared assets and resources standpoint.

In contrast, the second strategy embraces ambidexterity at the organizational resource level, creating a balance between core product lines and CI.

Supporters of this second path advocate for better alignment of innovation activities with the individuals who carry them out. They believe that enhanced CI performance comes from a deeper understanding of the skills and "attitudinal mindsets" required to perform CI's various roles. Ironically, proponents of organizational ambidexterity argue that CI should embrace more risk-taking and an entrepreneurial spirit, even though few individuals who choose to join large companies possess those traits. Regardless, consider this path. When aligned with the core business, it offers a useful approach for addressing the efficiency-change tug of war and participating in the Accelerating Change Race.

More Impact, Not Innovation

The third path builds on the first two. It advocates for creating large portfolios of initiatives. A larger CI portfolio is believed to increase the likelihood that at least one project will yield significant economic benefits. However, evidence from this "just do more" approach indicates that the results can be counterproductive. When applied across numerous companies over many decades, the law of large numbers suggests that the average economic value of a large portfolio will remain low. Unfortunately, it imposes more overhead on CI efforts. As a result, the large portfolio approach becomes more of a game of chance than a smart strategy.

Companies have tried and retried the three paths; yet, CI productivity still falls short. The state of CI explains why so many intelligent, independent thinkers in CI develop "entrepreneurship envy." If only I had the freedom that entrepreneurs enjoy, where pivoting is easier and success can be declared at lower revenue. These individuals find themselves caught between their desire for true independence outside the company and the economic security that a large organization provides.

The Best Practice Fallacy

Many academics and consultants advocate for greater CI productivity by analyzing what works and what doesn't. They filter practices based on CI results. However, we must recognize that the success rate of CI is extremely low, at around 3%. It's unreasonable to extract best practices from results at such low levels. We are delving into statistical noise. Any findings will yield only slight improvements.

CI requires much more than small, incremental gains. To ensure sustained support from boardrooms, shareholders, and executive teams, the gains must be at least four to five times the current base levels. While identifying best practices from existing methods may seem wise, it cannot deliver four to five times the gains.

If the slight impact of best practice research isn't enough to dissuade businesses from relying on it, this approach also targets traditional, single-product, orderly system thinking. This context, I have argued throughout this book, is flawed. Such best practice research resembles copying answers from someone else's test, only to find out that you are taking a different test.

Teams have implemented CI in the same manner for decades, and one might expect CI's effectiveness to be greater. Rather than refining what doesn't work, large companies need to transform their approach and reconsider their tools.

Many CI advocates assert that transforming an organization's culture and leadership will address CI challenges. While these improvements are essential, culture and leadership alone are insufficient to achieve the desired outcomes. A broader shift is necessary.

From Single to Multi-Product Thinking

Corporate innovation must undergo significant change, just like core product innovation and management. CI must transition from single-product orderly systems thinking to multi-product complex systems thinking. This transition demands a major shift in context, assumptions, tools, and processes.

The challenge is that most companies have integrated orderly systems thinking focused on single products into their CI operations and behaviors, as I explained in Chapters 5 and 6. This context hinders all CI initiatives. CI must transition toward a multi-product, complex systems thinking approach. It needs to foster practices, tools, and templates that support product line-centric initiatives. Most importantly, CI must create an environment that encourages contributors to explore the assembly and architecture of a product line system, as discussed in the previous chapters.

Corporate Innovation and The Accelerating Change Race

A large company's ability to navigate the Accelerating Change Race without effective change management remains problematic, even if it is possible. It's like playing a sport with one hand tied behind your back. Large companies often lack the ambidextrous finesse of the great Boston Celtics' Larry Bird, who, as a right-hander, famously shot left-handed for an entire game. Bird led his team to victory while scoring 47 points. To this day, no one has ever matched Bird's unbelievable feat. Don't bet on your company pulling off this feat in its innovation tug-of-war.

As long as CI remains ineffective, exploring radical innovations and pursuing notably unique markets and technologies will be out of reach for many organizations. CI must improve its effectiveness by focusing on the support and development of product lines as systems. Plus, as I discuss in the next chapter, it needs to collaborate closely with existing product lines, rather than working in isolation.

Chapter 27 Summary

Reorienting Corporate Innovation

- **Corporate Innovation's Current State:** CI aims to create new growth opportunities separate from a company's core products. However, it's ineffective, with most innovations failing to scale or succeed. That leaves CI groups with the challenge of explaining why they deserve to be funded even when financial returns don't warrant it.

- **Challenges and Misconceptions:** CI often promotes the false narrative that core businesses can't innovate, which pushes core product lines to focus solely on efficiency. These narratives harm the overall innovation, leading to a negative loop where neither CI nor core operations improve effectively.

- **Paths to Improving CI**: There are three common approaches to improving CI: isolating CI from core business (ring-fencing), integrating CI and core operations (ambidexterity), and increasing the number of CI initiatives. However, none of these paths have proven consistently successful, often because of misalignment with large corporation practices and methods.

- **Flawed CI Tools and Practices:** Existing CI tools and methods are inadequate, especially for handling the high levels of uncertainty and complexity involved. These tools are often too focused on single-product outcomes, neglecting the need for a more comprehensive, multi-product system approach.

- **Need for a New Approach:** To improve CI, companies must shift from a single-product thinking to a multi-product, system-oriented approach. This shift requires new tools, processes, and a reorientation of how CI initiatives are managed and executed.

- **Importance of Effective CI:** CI must become more effective for large companies to explore radical innovations and unique markets. This involves creating new product lines and integrating CI efforts with existing lines to keep pace in the Accelerating Change Race.

Chapter 28

Corporate Innovation's Main Job

Building Bespoke Product Line Systems

orporate Innovation (CI) in large companies must adopt a new approach to achieve significantly meaningful and successful outcomes. Engaging in more innovation theater and continuing to navigate uncertainty won't suffice. In this chapter, I explain why and how product line systems serve as the linchpin for successful CI. However, this chapter does not address what may occur or what CI teams must do when transitioning their work context to focus on building product line systems. Those topics are reserved for a different book.

The Big Shift

CI must shift its focus to advancing and transforming existing product lines before establishing new product line systems. This shift requires CI teams to view product lines as complex, multi-product systems.

The word "innovation" has prominently featured in CI's title for several decades. Over the years, many experts have argued that its role is to build new ventures and pursue breakthrough innovations. I agree with this, but I must emphasize that CI cannot fulfill its purpose without understanding and focusing on product lines. As I maintain, the job is to build and exploit product lines. Yes, commercializing novelty and breakthroughs, along with forming ventures, are desirable outcomes. However, CI, without a focus on product line as systems, creates an ineffective context for enabling the emergence of these desirable outcomes.

Corporate innovation must shift its focus from building, enhancing, and transforming product line systems to concentrating on new ventures and breakthrough innovations. This contextual change redirects corporate innovation towards multi-product complex systems, including their agents, components, constraints, and forces. It emphasizes product line stacks, system archetypes, business models, and ecosystems.

Product Line System-Centric

Integrating a product line system perspective into CI represents a significant "strategic initiative" for any business.[88] CI accomplishes this by collaborating with existing lines before developing new ones. This approach requires the involvement of CI, a product line team, and business leadership. CI works with a product line team, while top management guides, enables, and promotes the project.

CI activities shift when teams focus on developing and advancing "product line systems" rather than the usual emphasis on developing technologies, products, and businesses as "growth platforms." The product line system acts as the growth platform.

A CI team must first understand an existing product line as a complex system and determine its shortcomings and constraints in relation to both the present and future. They must work conceptually to form or reform the system's assembly of parts, enablers, and constructors. Their role is to explore and adjust the interplay of system agents and report one or more paths forward. The work of CI is not about establishing a business based on a single product or innovation.

The Traditional Approach

Consider how the standard CI approach seeks to transform technologies into products and products into businesses. Several shortcomings arise when we analyze this approach within the product line system context.

- Traditional CI often overlooks the need for a platform-lever. Without a platform-lever, potential systems may encounter challenges in establishing an architecture. This approach neglects the assembly, architecture, and system archetype of a product line..

- Verb-based segments are often missing in common CI initiatives. While CI efforts may use Jobs-to-be-Done research for a single product, I have not seen them segment markets based on multiple JTBD clusters, that is, verb-based segments. However, verb-based segments are a vital part of a multi-product stack. (See Chapter 9).

- Typical CI efforts lack a focus on creating an Outcome Velocity vector. Most CI teams ask top management not to judge their initiatives based on near-term cash flow. They also mistakenly believe that "Will you buy?" is the only question that predicts customer satisfaction. Plus, their efforts often overlook competitiveness until they deliver a product. There is no ongoing push to improve a line's Outcome Velocity vector. The perspective must integrate cash flow, customer satisfaction, and competitiveness at launch and after commercialization..

- You'll also see that CI efforts overlook the necessity of creating system enablers. However, a product line system without enablers is destined to fail. Teams must understand system enablers just as well as they understand products, technologies, and customer needs.

- Most CI managers recognize the importance of organizational fit but do not fully grasp chain-link alignment and its significance to a product line system architecture. They tend to react to mismatches during a product launch, often lamenting that the core business is inflexible. Throughout their development work, they fail to utilize the interplay among chain-links, the product line's archetype, and the business model. If you allow these critical influences to operate independently, you might as well disregard the concepts of competitive advantage and performance.

Unfortunately, these shortcomings kill CI's effectiveness.

Component Interplay

Building product line systems requires CI teams to explore and investigate potential system components and create variations of possible product line stacks. System components and stacks serve as the foundation for developing a system's architecture. This task involves identifying useful and impactful components and potential stacks while discarding those that are not beneficial.

There is no linear process for forming or reforming product lines. A step-by-step, linear process is nonsensical in the chaotic and complex uncertainty domains in which continuous improvement

(CI) must operate. Self-proclaimed CI gurus, however, commonly promote linear, step-wise approaches. While their selling pitch may sound intelligent, buying into it is not advisable.

To make matters worse, some gurus suggest that practitioners use tools better suited for orderly systems. From the perspective of complexity science, CI involves creating a complex system from a domain of chaotic uncertainty while refining the domain toward a complex state. Tools designed for orderly systems and linear processes operate in this environment only with a significant degree of luck.

A Dynamic, Fuzzy Puzzle

CI's job resembles assembling a puzzle without knowing what the final picture looks like, all while dealing with many oddly shaped pieces. Moreover, the reality is that the images cannot remain static; instead, they should evolve into a series of pictures, reminiscent of a movie rather than a portrait.

Just as there is no linear process for creating a product line system, there is no specific part or agent that the team must focus on to begin their work. You'll often hear that all innovation efforts should start by identifying customer needs. While incorporating knowledge of potential customers into the initial CI chaos can be beneficial, understanding specific customer needs at the onset of an initiative is crucial only when the product line system is already in place. You'll eventually require details about customer needs, but as a product line system develops, the Noun and Verb-based segments (key parts) will likely evolve.

Remember, Post-it Notes® did not emerge because of a customer's need.[89] You'll require numerous components and agents to develop a product line system. The sequence in which they arrive is not as crucial as whether CI teams encompass all of them. It will take several iterations of building and refining, constructing and deconstructing, to establish an impactful product line system.

A major CI failing stems from the belief that all you need is one product-in-the-market (PIM; see Chapter 9 and Appendix 7) to satisfy a single need set within one Noun-based market segment. Unfortunately, this reflects the single-product approach to innovation and product management, as well as the orderly systems way of thinking. In building product line systems, teams must create various types of parts and agents that are not accounted for in single-product thinking.

Learning & Auditing

If a company's CI unit cannot assist a product line team in enhancing an existing product line, it is unlikely that they will successfully create, develop, and launch a new product line system. Therefore, CI should initiate the transition to a product line-centric organization by learning PLSM and working to improve an existing line. I recommend that this first step precede applying PLSM principles to create a customized product line.

Introducing PLSM principles into a CI initiative can be disruptive. However, this new approach is justified because many CI initiatives do not succeed. A useful introduction can occur through a Product Line System audit. This audit assesses the work and outcomes involved in forming or supporting a product line as a complex system. Its purpose is to strengthen, not undermine, the initiative. It highlights what is working well and what isn't, identifies existing parts and agents, and indicates what needs to be created.

I recommend conducting an audit as part of the training for PLSM. This training-audit combination enhances product line systems learning. It also provides benefits sooner than waiting to implement PLSM principles until after CI teams believe they might have a unique product opportunity in hand.

Actions arising from a Product Line System audit will be situational. Each initiative is unique, and the effort to change, create, or remove a part or agent of the system will vary from case to case. However, developing the skills and knowledge to conduct the assessment is crucial for reorienting and reforming CI's common methods.

Work With Core Product Lines

The knowledge gained from advancing a core product line system can be invaluable. It enhances CI's understanding of product lines as complex systems and fosters insights specific to a core product line's current and potential parts and agents.

A strong interface with core product lines is essential to CI's contribution to a business unit. Contributions from core product lines should take precedence over efforts to achieve significant contributions to the corporation beyond a business unit. In other words, CI should support existing lines before developing new, independent product lines. It is important to remember that CI teams must adhere to a fundamental PLSM principle: Do not wedge a new product line archetype into an existing line with a different archetype. Instead, if such a change seems desirable, you'll need to learn how to "transform" the core line. Wedging only creates a mess.

I base the idea of initially focusing on core product lines on two facts. The rise or fall of current lines will have a significantly greater impact on a business in the near to mid-term than any CI initiative. Satisfying the desire to improve or transform (change archetypes) a core product line is more likely to be supported by CI's contributions than not. Furthermore, we know that working with existing components and agents is easier than creating a complete product line system from scratch.

CI teams must recognize that the behavior of a complex product line system is not just the sum of the individual behaviors of its components. When a CI team builds a system from the ground up, each additional part will change the system's behavior. CI work can frustrate teams until they develop the skills to identify and manage system behavior.

System Archetypes and Business Models

In Chapter 11, I discuss product line system archetypes and their relation to business models and product line strategy. It applies to CI just as it applies to existing core product lines.

Product Line Change Theory and Scaling

One of the most important steps in CI is scaling innovation to a level deemed essential for the overall business. However, scaling involves a type of change addressed by Product Line Change Theory. I discussed it in Chapter 22, and everything I mentioned there applies to scaling CI-created offerings. If a CI initiative seeks to scale an opportunity, it should plan and organize the effort with Change Theory in mind.

Corporate Venturing

Corporate venturing and corporate venture capital guide significant company investments in external startups. These investments often aim to establish complete product lines, but many also seek components and agents for existing or forthcoming product lines. From a product line system perspective, such investments can be crucial to CI success. A comprehensive understanding of PLSM can sharpen Corporate Venturing's investment focus and enhance its contribution to growth.

Creating Product Lines and The Accelerating Change Race

Corporate Innovation needs to define its role in addressing the efficiency-change dilemma. These groups should first focus on supporting changes and advancements to existing lines before developing bespoke product lines, distinct from the core business. For both roles, PLSM principles (the framework) offer CI the guidance needed to succeed.

CI's new role should not be to create innovations. Instead, it should focus on creating, facilitating, developing, and exploiting product lines as systems. While some may prefer to concentrate on scaling a business, establishing a product line system should come first, followed by a coherent stream of strategic moves. Scaling could be one such move.

Chapter 28 Summary

Creating Product Lines

- **Corporate Innovation Must Change**: CI needs to change how it works. Instead of just coming up with new ideas, CI teams should focus on improving existing product lines and creating new product line systems. This means learning about how different products work together as a system.

- **Change to Product Line Systems Building, Enhancing, and Transforming:** This new name better describes what CI teams should do. It's about building and improving product lines, not just developing new ideas.

- **Embrace Product Line as a Complex System:** Managing product lines as systems helps teams examine how different parts of a product line work together, including technology, products, and customers. The goal is to create or improve a whole system of products.

- **Stop Old Problems:** The old way of doing CI has problems. It often ignores important parts of creating a successful product line, like having a strong platform or understanding customer needs. These issues can make it hard for new products to succeed.

- **Emergence is not a Predetermined Outcome:** Creating a product line system is like putting together a puzzle without knowing what the final picture will look like. There's no step-by-step guide to follow. CI teams must try different combinations of parts and be ready to change their plans as they learn more.

- **Focus on Product Line Creation:** CI teams should start by helping improve existing product lines before trying to create new ones. This approach helps them learn about product line systems and how they work. It also allows CI to make a bigger impact on the business right away.

SECTION SEVEN

Amping Up Gains and Anchoring Our Thinking

"Theory without practice is empty and practice without theory is blind."

– Immanuel Kant, circa 1780; Philosopher

Fusing data, ML, and AI will enable previously impossible platform-levers and agent constructors. These new system components will transform every industry. Those companies that don't change their offerings to match industry and market changes won't survive. We must work together to understand and benefit from this next era of product innovation and management.

- The Author

Chapter 29

Amplifying and Accelerating Product Line Systems

AI, Intense Focus, Flow Resonance, Coherence, and Venturing

Historically, improving a product line's performance required teams to seek greater efficiency in its components, particularly operational technologies. Today, things are very different. The combination of AI, ML, domain-specific knowledge, and data gathering, along with AI reasoning and planning, flow automation, and system component creation, has the potential to significantly enhance a line's efficiency and expand its Change Capacity.[90]

Experts tell us that AI is increasing its power by three to five times each year, a trend expected to continue.[91] Many argue it is likely to accelerate as AI gains the ability for self-improvement and as quantum computing capabilities develop. This implies that AI could surpass its current power by over one million times within a decade. Such AI power will amplify PLSM well beyond the performance provided by single-product orderly systems thinking. It will only be limited by human and societal constraints, rather than technological advancements.

Our accumulated knowledge and decades of experience in product innovation and management have served us well. However, the research, teachings, methods, and practices we've embraced act as the prologue to the epic that is about to unfold. It is both exhilarating and frightening to see what lies ahead.

> *At the time of this publication, most organizations are exploring the use of AI and AI agents in product innovation and management. However, they primarily focus on cost reduction and speeding up their approach for single products and structured systems. The larger benefits will come from leveraging AI to enhance the complex multi-product systems approach. This involves creating enablers, key parts, flow constructors, and ecosystems, while also facilitating the interactions and interdependencies of these agents. Ultimately, the goal is to devise and execute a coherent stream of strategic moves that drive compounding gains.*

From Integration to Modelling

In the early 2000s, the German enterprise software company SAP engaged me to support their teams working on product lifecycle management software. This experience was enlightening. Integrating data and applications is far more powerful for product innovation and management than the homegrown, standalone software that most companies were using. I witnessed firsthand how integrated data and software applications enabled organizations to enhance their efforts in ways that were previously impossible. I first wrote about

"Integration-enabled" product innovation and management in 2005.[89]

In 2018, researchers at MIT published a paper entitled "Engineering with a Digital Thread."[93] The paper focused on design engineering and the use of Digital Twins. The concept of the Digital Thread aims to integrate all product innovation and management information, facilitating the creation of a digital model, or a digital twin, of the product, its development, and its manufacturing.

The Digital Thread aims to integrate all data across ecosystems. However, the current approach focuses exclusively on enhancing individual products, supply chains, and finances. It has yet to expand to the entire product line system, but that will change.

As teams begin to supercharge their product lines, they must adopt a guiding strategy. This enabling constraint establishes boundaries for integrating automation, machine learning, and data analytics with AI, or, as I refer to it, "AI enablement."

AI-Enabled Everything

Today, Robert Cooper, a product development expert who has made significant contributions to the field, is actively exploring and showcasing AI-enabled product development.[94] We also see that every product development support software vendor, from PTC to Planisware and Atlassian, is working to leverage AI to enhance their tools and provide greater benefits.

Any task, whether cognitive or physical, can be assisted or replaced by AI agents. Think of AI agents as applications that process data and information, then, using their coded capabilities and interfacing with an AI model, return text, numbers, or images. When agents are purposefully linked in a flow and the output is refined or expanded, the results can be impressive. The challenges teams face in creating such integrated agent support are understanding the flow, developing the code, and selecting and building the AI models. However, AI can aid in the creation and implementation of solutions to address each of these issues.

From my perspective, understanding product lines as multi-product complex systems is the most significant hurdle for companies. When AI initiatives focus on product development and innovation management solely from the perspective of a single-product orderly system, the deliverables will fall far short of AI's potential. In such cases, there will be no gain from innovation charters Verb-based segments, platform-levers, and cross-product attribute positioning.

To be clear,

The only way to fully exploit AI in product innovation and management is to manage product lines as complex, multi-product systems.

If you exploit AI while remaining in the single-product orderly system paradigm, you'll be turning up the spigot on BS products, as I discussed in Chapter 2. You'll forfeit the opportunity to gain from platform-lever and cross-product attribute positioning, both advantages that will grow in importance as companies seek to keep pace in the Accelerating Change Race.

Gains Won't be Overnight

Consider the comparison of AI enablement to digital transformation, the change initiative that nearly every company has undertaken over the last few decades. Major gains were achieved only after companies integrated organizational changes into their digital rollout.

In product innovation and management, we must incorporate fundamental changes to both the product offerings and the ways we manage products to realize significant gains from AI enablement. These changes go beyond simply supporting existing workflows; instead, they tackle tasks that were previously impossible.

AI has become an increasingly essential component of product offerings. The key advantage in product innovation and management lies in the impressive power and agility that AI-enabled PLSM can offer to each company.

It's Already Begun

The reason for PLSM's superiority lies in how the framework encourages teams to focus on the interplay of the system's components and agents. This is evident when examining the roles of the product line roadmap and the system assemblage map. These system enablers do not exist within a single-product orderly system context. In the context of a complex, multi-product system, AI-enabled systems can help us understand and influence the interactions among the product line system components and agents in ways that were previously impossible. The "can and can't" circumstances change dramatically.

PLSM establishes a framework to enhance the performance of system agents and components, particularly those related to platform-levers and constructors. Once again, this is a game-changer.

Platform-lever Courage

Since Henry Ford introduced the production line, customers worldwide have benefited from this extraordinary platform-lever. For decades, production lines were the only type of platform-lever widely used. Fast forward to recent years, and now many types of platform-levers deliver superb bang-for-the-buck performance for thousands of product lines. We now see various technologies, such as software operating systems, automated production, and AI-driven self-service, integrated into platform-levers. See Appendix 6.

The two impressive gains we see with platform-levers are development speed and economies of scale, both of which deliver greater value for the investment. That's why the level of investment companies have made in strategic initiatives related to platform-levers has grown remarkably.

Consider Tesla's $5 billion battery plant in Nevada, announced in 2013. At that time, Tesla's sales were only around $2 billion. This represented a monumental platform-lever move.

Remember the Archetype Disruptor Theory that I laid out in Chapter 12. When platform-levers are new to an industry, as was the case with Tesla's battery and its production, they have the potential to disrupt that industry. That was the gamble. Panasonic felt confident about the move, agreeing to invest between $1.5 and $2 billion, while Tesla covered the remaining $3 billion. Tesla's behavior resembled that of a startup, but the difference was that Tesla's financial statements included a few extra digits at the end of the figures.

Or consider Apple's platform-lever move to design and manufacture its own CPUs, a development that dealt a near-fatal blow to the mighty Intel. These significant strategic actions were once viewed as too risky. It takes courage to implement them. However, it now seems overly risky to avoid major platform-leveraged initiatives. A single company's bold platform-lever move, whether AI-enabled or not, can spell doom for another.

Remember, too, that embedding AI into a platform-lever will change a system's archetype. This alone can be a potent strategic move that disrupts an industry.

One of the difficulties for many of us who grew up observing long-established product lines is that it leaves us expecting new platform-levers, particularly those that enable industry disruptions, to have lengthy life cycles. However, that's not the case; it's almost the opposite. If you've identified a platform-lever that is new to your industry, you should anticipate that others will quickly follow with similar platform-levers or attempt to leapfrog you with an even better platform-lever. BYD did this to Tesla, albeit with the help of the Chinese government.

Tesla's strategic moves suggest that they aim to outpace and out-invest competitors in developing entirely new platform-levers. However, notice how other automotive companies have partnered with battery manufacturers to expand their ecosystem and allow them to outperform Tesla's battery performance.

Meanwhile, Tesla also seems ready to join others with newer batteries while raising the stakes with autonomous driving software and gigacasting integrated into automated car assembly—a fundamentally new platform-lever mix and system archetype.

The Constructor Battle

AI can serve as a constructor for building other system components or replacing specific processes and flows within a system. This powerful agent underscores the importance of comprehending the assembly and architecture of product line systems, as well as the various flows within these systems.

When properly defined and administered, constructors are the most powerful enablers of any product line system agent. Consider what Amazon's CEO, Andy Jassy, revealed in a LinkedIn post in August 2024. He declared that AI saved the company $250 million and 4,500 person-years of software developer time.[95] Jassy said AI was conducting massive coding at a low cost and nearly instantly. In PLSM, we refer to this as a "constructor." Amazon's AI tools set is "constructing" workflows that create system parts and do so incredibly fast. The constructor builds the agent to do the work, but remains independent of the system.

Indeed, many traditionalists may choose to criticize the quality of AI-influenced outcomes by claiming that AI produces low-quality results and hallucinations. They argue that AI hinders people's abilities to imagine, theorize, or rely on intuition. However, this stage of AI's development is still in its infancy, and its reasoning and learning capabilities are limited. Sooner or later, AI-enabled Constructors will impact every aspect of product innovation and management, including customer-facing concept generation, planning and roadmapping, software and hardware development, as well as marketing and sales. Don't be surprised when an industry player who speaks negatively about AI quickly turns to exploit it.

The shift toward building complex system components and enabling work, decision-making, and information flows with AI is significant. Tesla's DOJO computer exemplifies the essential role of constructors. This homegrown supercomputer and AI are intentionally designed to code and continuously enhance the company's autonomous driving algorithm platform-lever. Dojo is neither a platform-lever nor a product.

Enabling Constraints, Flow, and Prompt Engineering

When companies transition into PLSM, one of the first tasks they need to tackle is setting up enabling constraints. These agents are crucial for managing complex systems. The two enabling constraints that are most helpful in product line management are innovation charters and strategy. (See Chapter 15.) Each is also a starting point for creating AI-interfacing prompts.

Suppose a potential innovation charter is being considered for placement on the roadmap. Right away, you might use the charter in a prompt that requests an AI model to come up with, let's say, twenty concepts that satisfy it. The model may also be queried for an analysis of the constraints embodied in the charter, as well as helping to identify constraints that are not reflected in it.

If the AI model is an LLM paired with industry, company, product line, technology, and customer-specific RAGs (see Figure 29.1), the prompt results are likely to be insightful. Take this a step further. Allow the model time to reiterate the job to minimize irrelevant results and highlight potentially valuable products or platform-levers. Then, add human interactive dialogue that further explores constraints and potential results.

From Prompts to Constructors

Now, integrate this prompt-initiated flow into roadmapping, project portfolio management, and review it considering the line's strategy (another enabling constraint). Within this extended flow, how might the AI model perceive the potential key components on the roadmap as projects within the portfolio of ongoing projects, and in alignment with the line's strategy? What rationale narrative could the AI model provide to suggest that the potential product or platform-lever will enhance the Outcome Velocity of the line?

An AI application that facilitates the creation and enhancement of flow through insights and judgments would be a Constructor. And, if this Constructor proved proficient, its value would be immense. No company or product line team would want to be without it, especially when competitors have one.

No doubt, creating such a Constructor would be costly. But, as I expect, future capital investments in product innovation and management will emphasize both platform-levers and Constructors. That's how the PLSM game will be played as company after company seeks to outpace the pack in the Accelerating Change Race.

Our Natural Bias

I'd be remiss if I didn't express my concerns about the Dunning-Kruger effect[96]. This cognitive bias reveals that people with limited expertise in a topic tend to overestimate its potential when first learning about it. That describes me and AI. It also applies to the AI experts and PLSM. With that caveat in mind, let me share how other approaches utilizing AI enablement might impact product line systems and their behaviors. Table 29.1 below explores the potential of AI enablement that can support and drive PLSM.

Table 29.1 Possible AI Enablement

PLSM Element	Value-Adding AI/ML-enablement
System Modelling, Assemblage Mapping, Behavior Monitoring,	Facilitating the system, suggesting system issues and opportunities, and revealing behavior patterns/changes

PLSM Element	Value-Adding AI/ML-enablement
Outcome Velocity Reporting	
Platform-levers Design and Maximization	Analyzing the system, tracking leverage, organizational focus, and multi-platform mix coherence, facilitating insight into redesign and archetype transformation.
Constructor Creation	Using AI-enabled constructors to create or replace flows, build parts, and boost other enablers.
Verb-based Segmentation	Facilitating Jobs-to-be-Done research and outcome clustering; Analyzing alternative segmentation schemas. Optimizing platform-levers based on cross-Verb-based segment analysis.
Noun-based Segmentation	Facilitating research and creation, analyzing segments, and suggesting and rationalizing priorities.
System Stack and Product Line Roadmap	Creating, tracking, and analyzing a line's key roadmap; Suggesting innovation charters and strategic move priorities.
Flow Automation and Resonance	Identifying orderly flows that can be automated. Assessing the coherence and resonance of these flows.
System Constraints and Forces	Tracking and monitoring system constraints and forces.
Weak Signal Sensing	Aiding teams with general and specific information handling, probing and compiling global narratives, shaping and evaluating information.
Centers of Technical Excellence	Aiding these organizational entities with up-to-the-minute information, compiling specific and broad public knowledge
Strategic Move Formation and Coherency Evaluation	Helping evaluate and refine or reshape strategic moves and coherence in light of system constraints.

PLSM Element	Value-Adding AI/ML-enablement
Composability	Aiding teams in identifying and creating ready-made building blocks to add value to platform-levers as niche or customized products.
Insight Generation & Sensemaking	Aid team members to make sense and create coherence with the system
System Enabler Improvement; System Assemblage Mapping	Performing analysis of enablers and potential improvements, prioritizing enabler creation, changes, and improvement, mapping, and prioritizing enabler advancements
System Change Capacity	Aiding teams in evaluating agent Interplay and their ease of change, offering up potential change
Chain-link Alignment Assessment	Actively analyze and report chain-link alignment specific to system behaviors and flows; Suggest chain-link alignment / boost improvements.
External Ecosystem Assessment	Actively analyzing and reporting ecosystem alignment specific to system behaviors and flows; Suggesting chain-link alignment / boosting improvements.
Distributed Initiative and Leadership Support	Actively coach and guide distributed leadership roles, facilitating individual initiatives and aiding in project content and management.
Product line system governance	Dashboard monitoring and reporting Outcome Velocity, initiative, and project reporting
PLSM Principles	Maintaining, improving, and aiding the communication of all principles

PLSM Principles

Don't expect integrating AI into product line innovation and management to be easy. It won't be. AI technology is evolving rapidly, so what you implement today will likely become outdated soon. Nevertheless, every product line team must embrace this challenge.

Individuals or ad hoc teams should be encouraged to select one or more of the elements. As part of insight generation and sense-making, their role is to determine how AI might enhance the contributions of these elements to the line's performance or increase capacity. There are many possibilities. Generally, teams should evaluate these possibilities by weighing the energy required to make AI enablement a reality against the expected benefits of successful AI enablement.

The team should pursue a portfolio of AI enablement rather than concentrating on a single instance. If PLSM is new to your organization, researching how AI can enhance an element's contribution may be beneficial for both understanding PLSM and adding value to the system.

Adding Reasoning and Learning: Beyond LLMs and RAGs

In the past few years, AI scientists and engineers have diligently worked to develop knowledge models known as "World Models" that emulate how people understand the world around them. These are predictive models (moving from one state to another) that reflect the dynamics of a specific topic area. PLSM establishes a framework for a world model of multi-product, complex product line systems.

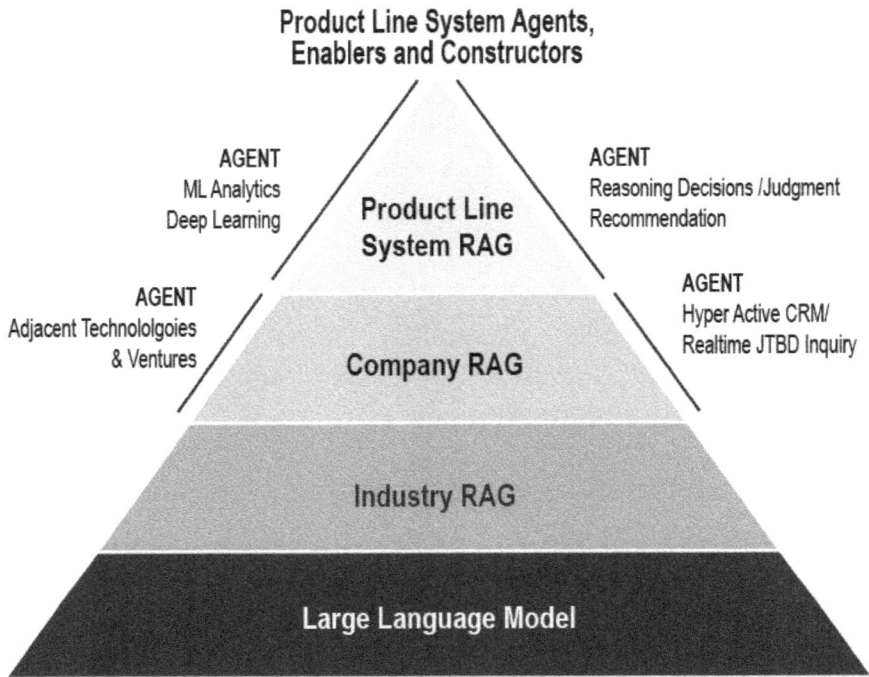

Figure 29.1 Product line systems demand an AI stack specific to the job.
Everyone rents the LLM foundation for their AI efforts. The question big companies face is, "How good, in all dimensions, are your industry, company, and product line Retrieval-Augmented Generation modules?" And, "How good are the system and AI agents you use to effectuate the Product Line System AI Stack?"

The significant advancement in world models has been the integration of neuro-symbolic reasoning. Consider this technology as enhancing knowledge models through reasoning and learning. These systems identify, explain, and analyze complex patterns, and then outline the rationale for potential actions while continuously learning from what is effective and what is not.

AI enablement will evolve through a composite layering of tools. It will start with LLMs, to which teams will add:

+ Industry/company/product line RAGs

+ PLSM knowledge (world) models

+ Neuro-symbolic reasoning, learning models, and context memory

+ Human Interface specific to ethics, paradox understanding, trust, and decision coherence

+ Job, task, process, and purpose-driven action-taking agents,

+ Abridgement between reasoning, goal setting, planning, and system scenario testing

+ Real-time environmental signal analysis, responsiveness, and resiliency, to create a

→ **A recursively self-improving PLSM model-constructor that accelerates planning and strategic moves... radically and continually improving the line's outcome velocity.**

This last point is the holy grail of product innovation and management. I view this model-constructor as a multi-trillion-dollar opportunity for product innovation and management. And while I haven't discussed it, the scenario that prevents us from reaching this point is if or when artificial superintelligence outruns society and forces the end of capitalist democracy. That, however, is for another book.

Amplifying Weak Signals, Insights, and Sensemaking

In chapters 17 and 18, I emphasized the importance of insight generation and sensemaking in shaping a powerful stream of strategic moves. PLSM requires a deliberate and dedicated focus on these tasks. Not only must the affordance discussed in Chapter

18 be part of this focus, but someone or some team must also take responsibility for continuously improving the enablers of insight generation and sensemaking. This work must be integrated into the system's assemblage map.

Every product line team should prioritize using AI to sort information and identify weak signals. Understanding these signals is crucial for strategic moves. The better a team understands information signals, the more capable they are of making impactful strategic moves. AI will enhance the signals and help filter what is important from what is not.

We also know that the speed and quality of insight generation and sensemaking are the most important skills a team can develop; this skill focuses on conjuring and executing safe-to-fail experiments. I am unaware of any approach that utilizes AI to identify and facilitate such experiments. Yet, someone, some company, or some service will undoubtedly master this capability. When they do, every product line team should leverage it.

New Ecosystems: PLSM Venturing

A product line's external ecosystem can be essential for driving product line success, regardless of whether AI is utilized. You don't need to develop your system components and agents in-house.

Partnering, venturing, and sourcing can have a more significant impact on the system. However, this necessitates an open approach to permitting outsiders to engage with your roadmap and assembly map. Apple could not have pulled off its Apple Silicon moves without involving the massive chip manufacturer TSMC.[94]

A product line's external ecosystem can provide tremendous value. Paul Hobcraft has extensively researched and written about external ecosystems.[53] To quote Hobcraft, "Ecosystem design and thinking can address more complex customer needs and global challenges, leverage technological advancements, and recognize that a more fluid set of boundaries between organizations is blurring or vanishing far quicker today."

Intentionally designed ecosystems enhance product line performance. When your line's ecosystem outshines a competitor's ecosystem, it can create a substantial competitive advantage, increase customer satisfaction, and lead to higher contribution margins and positive cash flows.

Amped-Up Enablers

It's unlikely that solution providers will change their single-product orientation without demands from their customers. Companies with the greatest need for such aid will take the first steps in enhancing multi-product systems enablers. They'll be the lead users and innovators. Tesla declared that role as if saying, "We can't wait for someone else to create DOJO; we need to build this enabler ourselves."

Vendors of a software enabler package should embrace the lead user theory[98] and focus on the lead users of enablers, rather than on their larger customers. In my experience, addressing an initial need for software that enhances a product line enabler is more likely to be met by a new startup than by an established software company.

Environmental Boosts: Culture, Leadership, Affordance

Not all PLSM supercharging needs to be driven by AI. Culture, organizational structure, and leadership influence how people behave and are critical factors in enhancing a line's performance and Change Capacity.

Numerous studies have examined the effectiveness of these environmental enablers. However, they primarily focus on single-product systems. Once again, a significant distinction exists between system enablers for orderly systems and those designed for complex systems with multiple products. This distinction is critical.

New Capability, New Capacity

Creating and executing a series of impactful strategic moves is one of the biggest challenges in product line management. It's like solving a puzzle with fuzzy pieces and a limited understanding of what the final picture might look like. Each insight into the product line system is a puzzle piece, and the picture is always somewhat uncertain.

The primary role of product line management is to create and coordinate strategic moves that, when interwoven, influence the line's Outcome Velocity far more than just the sum of the individual moves. Supporting this role significantly enhances the management of product line systems.

Teams accomplish this task using Product Line Roadmapping. It's an innovative approach—a new enabler—to support product line systems management. Instead of merely displaying project timelines on a graph, it serves as the primary control center for the product line. It assists in gathering information, making decisions, and understanding how everything aligns. Think of it as the system's targeting and planning tool.

Many topics converge in Product Line Roadmapping. Its primary purpose is to help teams translate insights into actions and make informed decisions. You can read more about product line roadmapping in the online appendix. Most product line teams would welcome AI-enabled roadmapping and assembly mapping for their product lines.

Much More to Come

In the next chapter, I examine other developments and fundamental insights that the product innovation and management community needs to address to gain more advantages from PLSM.

Amplifying and Accelerating Product Line Systems to Blow Past Competitors in The Accelerating Change Race

The potential to enhance and expedite product line systems using PLSM self-learning knowledge models, along with AI-enabled flow constructors, automation, insight generation, and sensemaking, is substantial. This upcoming change will disrupt industries and companies on a large scale.

Advancing product lines requires two key system enablers: assemblage and key part roadmapping, which must play crucial roles in product line management. Without these enablers, system coherence will be compromised.

Imagine a consortium of large companies, each with multiple product lines, pooling their resources and expertise to secure government or investor funding for the development of a massive AI-enabled PLSM Constructor. The total investment would surpass $1 billion. They aim to create powerful AI-driven constructors capable of generating system agents such as technological building blocks, work and decision flow enhancers, weak signal sensors, and assembly mapping tools. These companies would then align their product lines to

417

consistently disrupt industries with new, more advanced product line system archetypes.

Change, not efficiency, will dominate the strategies of companies that embrace supercharging their product line systems. Unfortunately, those product lines that fail to adopt this approach, I'm afraid, will require exit strategies.

Chapter 29 Summary

Amplifying and Accelerating Product Line Systems

- **Focus and AI-enablement:** Companies can supercharge their product lines in two primary ways: by focusing intensely on optimizing the interplay between different components and by leveraging AI to explore and refine the system. The best approach is to use both methods along with coherent strategic thinking.

- **All System Agents and Parts:** AI is transforming how companies manage their product lines. It's not just about making one product better but about improving entire systems of products. Product Line Systems Management (PLSM) is significantly more powerful than traditional methods.

- **Making Bold Moves:** Big companies can make bold moves with their product lines. For example, Tesla spent $5 billion on a battery plant when their sales were only $2 billion. These major shifts, often involving new technologies or AI, can completely transform how an industry operates.

- **AI and System Constructors:** Constructors can help teams build things more efficiently and cost-effectively. Amazon's CEO stated that AI has saved the company $250 million and 4,500 years of work time. This is changing how companies develop products and will render some existing working methods less important or obsolete.

- **Both Products and The System:** AI enablement can help with many aspects of product line management. It can help design new products, gain a better understanding of customers, and identify new opportunities. Companies should experiment with using AI in different ways to determine what works best for them.

- **Boosting Performance with Ecosystems:** Product line teams should collaborate with other companies and seek outside help to improve their products. This might mean sharing plans with partners or using new software tools. The best product lines will be those that can work well with others and use new technologies smartly.

Chapter 30

PLSM Praxis: Theory and Practice

A Call to Academics, Consultants, Trainers, Software Vendors, and Practitioners

---

This book presents a novel perspective on product innovation and management. However, please recognize that this perspective, which emphasizes product lines as complex systems, is neither comprehensive nor final. There is still much more we need to learn.

The significance of managing product lines as systems is too great to overlook the need for a more comprehensive understanding that incorporates diverse viewpoints. Future cases and narratives will shape and provide context for the framework, and enhanced theories and practices will lead to improved principles.

I strongly encourage counterarguments to my emphasis on multi-product systems instead of today's prevalent single-product approach. I advocate the same view from my perspective on product lines, considering them complex rather than orderly systems. All academics, consultants, practitioners, and software vendors who believe they must continue with a single-product, orderly systems perspective in product innovation and management should help me and the community understand their position.

Changing Thought Worlds

In 1982, I joined a new professional community called The Product Development and Management Association (PDMA).[99] At that time, its membership was primarily academic. Many of the early members aimed to build careers in teaching product development. PDMA, through its Journal of Product Innovation Management and research conferences, has effectively supported the academic community while sharing knowledge and insights with practitioners.

At one of our early meetings, I asked two of the association's founders, Bob Rothberg from Rutgers University and Merle Crawford from the University of Michigan, why they included "and management" in the association's name when everyone focused solely on "development." I found the answer intriguing. It turns out that no academic was teaching, and only a few were researching product development. For any academic looking to pursue product development and innovation, acknowledging product management was essential to gain support from the heads of their marketing departments. Thus, product management secured its place in PDMA's name.

For years, "development" coupled with innovation has been PDMA's core topic, leaving "management" of existing products to other groups, associations, and academic publications. Now, though, these topics cannot be separated. They are intertwined when we focus on product line systems. One influences the other. The outcomes that emerge after products enter markets are inextricably linked to the what, how, and why of their development.

We have advanced well beyond the initial support that PDMA offered for product development. Today, a new world emerges as we embrace a complex systems perspective on product lines. The performance of a line is heavily influenced by how effectively an organization navigates and manages the constraints, key parts, enablers, constructors, and forces of its product line system. The PLSM framework is quite helpful, but we require more.

Product Innovation and Management Praxis

A significant issue for traditional, old-school product innovation and management is its "praxis." Praxis is the combination of theory and practice. My perspective over the last four decades is that product development and innovation management have been rich in practice but lacking in theory. To paraphrase a philosopher somewhere, "Practice without theory is to sail without a rudder. Theory without practice is not to set sail at all."

I believe traditional product development based on single-product thinking is rudderless.

Consider a typical academic study in product development. The research examines outcomes related to practices at various companies by dividing performance into the top, middle, and bottom thirds. Researchers then analyze the data to understand which practices differed for the top third compared to the bottom third of outcomes. This difference delineates the "best practices" relative to the outcomes.

Theories do not drive traditional product development. Instead, practitioners devise an approach or method; academics assess whether it is "best," and consultants and software vendors sell the best practices back to more practitioners. After many decades, this symbiotic relationship has pushed single-product orderly systems into every company.

Theory Staying Power

Only a handful of theories related to product development, management, and innovation management appear to have lasting relevance. For instance, consider the Jobs-to-be-Done theory, Disruptive Innovation, Time-based Competition (time-to-market), and Lead-user theory. Each addresses a segment of the product development field, but none helps to integrate these segments. Unsurprisingly, each is rooted in single-product thinking. These theories were not developed from fundamental principles in the general sciences.

Complexity theories apply across various substrate types, including environmental, biological, sociological, and economic. A behavior of complex systems that only aligns with the multi-product substrate would likely be a subtlety of that substrate and insufficient to challenge the complexity science theory. Nevertheless, this subtlety would be very helpful to understand.

It is crucial to develop and refine theories for complex multi-product systems. We cannot rely on advancing practices without a solid theoretical foundation.

The major problem with the old-school "best practice, theory-less" approach has been the time it takes to advance practices. It can take up to a decade for a practice to establish itself in multiple companies, followed by a few more years to research, analyze, and share that practice, and then additional years for other companies to adopt it as their own. By the time the practice is fully implemented, the originating companies have likely moved on to new and different practices. This approach is mindless when, as Stuart Kaufman says, we are in an exponential upswing of adjacent possibles.

We should not rely on outdated best practices as we transition to managing complex multi-product systems. Instead, we need a modern foundation built on well-established theories and sensible practices, i.e., the praxis. We are fortunate that complexity science and systems thinking have provided us with solid theories and principles over the past fifty years. Our goal is to understand and utilize these concepts in product line management and to create a robust praxis for managing product line systems.

New theories and practices must emerge. Our challenge is to stop depending on the outdated method of measuring and replicating best practices. Instead, we must think our way forward.

Academic Call to Action

My call to academics is to conduct both secondary and primary research to argue for or against the shift towards product lines instead of single products, and to complex systems rather than orderly systems.

Consultants and software providers are welcome to wait a decade or more for academics to sort through the foundations of PLSM. However, do so at your own risk. As with most innovations, product development, and management issues, academics do not lead the way; it is practitioners and consultants who do.

Practitioners and consultants promote the early adoption of practices and methods. We have seen this consistently. Consider Agile and Project Portfolio Management; they set the stage for each approach. Academics then evaluate and refine these methods, often rebranding them.

You'll find the same for Jobs-to-be-Done and Design Thinking Theory. While we consider these to have an academic origin, those of us who've been in the field for many decades recognize their practitioner roots. Jobs-to-be-Done draws from Tony Ulrich's Outcome-Driven Innovation[100], and Design theory builds upon the work of Alex Osborn[101] and various planned innovation programs[102] offered in the 1980s. Plus, it's no coincidence that academics who propose a theory maintain significant consulting practices that accompany it.

Academic support for a theory gives it much greater credibility. Their work enhances the theory by bridging the adoption chasm, as described by Geoffrey Moore in his book "Crossing the Chasm."[103]

Academics are well-equipped to formulate, test, refine, and reshape foundational theories, corollaries, and principles. This is no small task, as it requires delving into the theoretical aspects of complex systems, knowledge management, strategy, and innovation—all of which are currently evolving and adapting to our world. Specifically, I encourage leading innovation scholars to collaborate with prominent scholars in complexity theory to address this challenge. Historically, though, academic cross-disciplinary teaming has proven more difficult than cross-functional teaming in business.

Table 30.1 below presents several theories and models that should be reconsidered from the perspective of complex, multi-product systems. For example, I have shared my insights on archetype and business model alignment (Product Line Congruency Theory), on changes to Radical and Disruptive Innovation (Archetype Disruptor Theory in Chapter 12) and on Time-based strategy (Time-to-Market versus the Outcome Velocity Vector in Chapter 4 and Appendix 4). I also included Product Line Change Theory in the discussion.

Table 30.1 Theories, Laws, Models, & Frameworks

Theory of Constraints - Eli Goldratt	McKinsey & Company's Three Horizons
Jobs-to-be-Done Theory - Clayton Christensen	Customer Loyalty Model - Richard Oliver
Organizational Contingency Theory - Fred Edward Fiedler	Time-based Competition - George Stalk (BCG)
Lead User Theory - Eric Von Hippel	Blue Ocean Strategy - Chan Kim & Renée Mauborgne
Radical Innovation Framework - Leifer, McDermott, O'Connor	Quality Function Deployment Framework - Yoji Akao
Disruptor Theory - Clayton Christensen	Balanced Scorecard - David Norton
Diffusion of Innovation Theory - Everett Rogers	Market Share / Growth Matrix Model - BCG, Directional Policy
Wright's Law - Theodore Wright	McKinsey/GE Matrix Model: Invest/Protect/Harvest/Divest
Learning Curve - Hermann Ebbinghaus	Decision theory | Think Fast and Slow - Daniel Kahneman
Business Model Theory - Alexander Osterwalder, Yves Pigneur	Leverage and Scale Competition - Panzar and Willig
Five Competitive Forces - Michael Porter	Where to Play/How to Play - Roger Martin
Diseconomies of Scale: Scale makes change harder - Fisher Body Co./GM	First Principles Thinking: start with established science - Aristotle

Evaluating theories, laws, models, and frameworks against a product line's complex and diverse system might change our understanding of both the guiding principles and the system itself. If the theories and models stay the same, we need to delve deeper to uncover why and integrate that knowledge into how we manage product lines as systems.

Theories behind complexity science and complex systems originate from various fields, including biology, physics, and sociology. It's important to recognize that the complexity theory of social systems, such as business, differs from environmental and biological systems. Furthermore, we need skills in knowledge management, learning, and strategic thinking for effective product line management.

Practices and Tools Added to Theories

In the 1980s, many companies began using basic tools like Microsoft Excel to improve their processes and practices. Experts referred to Excel as a "point solution." In the early 2000s, more comprehensive solutions became available for project, process, and portfolio management. Yet, decades later, these enabling tools still remain oriented toward a single-product, orderly system way of thinking.

Today, there is little doubt that the product development practices of large companies are superior to those we followed a few decades ago.

As we advance in managing complex multi-product systems, we also need to develop and implement enablers, tools, and constructors to support the growth, enhancement, and adaptability of our product line. I recommend that practitioners, service providers, and consultants work together to create enabling tools, templates, and processes.

Every product line system includes numerous agents, parts, and flows (see Chapter 24). Enabling practices can support or improve each one. An interesting characteristic, however, is that each tool's functionality is tailored to the product line system's archetype.

I highly recommend that practitioners collaborate with others who manage product line systems of comparable size and type, but serve different industries.

Table 30.2 below presents key system components and practices that offer opportunities for improvement. Executing this work can integrate practices into theory, refining the approach to enhance the PLSM praxis.

Table 30.2 Potential Practice Improvements via Software

Innovation Charter Template / Creator	PLSM Dashboard / Communicator
Assemblage Mapping Tools (Stacks & Change Maps; Consider Cynefin Co's Hexi Tool Set)	Outcome Velocity Future Caster & Tracker
Stack Maker & Product Line Roadmapping Tools	Decision / Judgment PLSM Principle Alignment
Insight Generating, Building, & Tracking	Orderly System Root Cause Analysis /Roadblock Toolkit
Sensemaking Tools (Consider Cynefin Co's Sensemaker Tool)	Scenario Back-Casting Tool
Insight Generation And Sensemaking Affordance Manager Signal / Weak Signal Detection	Change Capacity Evaluator PLSM Tracker
Strategic Move Formulator/Planner: Energy Vs Time	Constraint Mapper / Type Designator; Energy, Time, Optionality Mapper
Constructor Framing, Blueprinting, Planning, & Building	Ecosystem Coordinator / Communicator

Flow Enabler & Resonance Evaluator (Decision, Work, Information Flows)	Archetype - Business Model - Business Strategy Evaluator & Monitor
Alpha-Omega Product Portfolio Management (Includes Innovation Charters & Product Retirements)	Platform-Lever Monitor Health, Factors: Speed of Development & Bang For The Buck

What's Needed

We need two significant changes to take hold. First, leaders in large companies must recognize the importance of integrating change into their core product lines. This does not detract from pursuing major innovations outside of core offerings. Instead, it acknowledges and responds to the reality that technologies and customer needs change too quickly to focus exclusively on efficiency and assume that it is sufficient.

The second change involves everyone, including both academics and practitioners, recognizing the critical role of product lines and their behavior as complex systems. Pushing harder on traditional single-product and orderly systems approach to product innovation and management has run its course.

What's Next

Those involved in product innovation and management who recognize the significant advantage PLSM offers to companies should establish and implement enablers and constructors similar to those listed in Table 291. Such enabling practices and tools represent new territory and necessitate considerable ingenuity. For instance, consider creating system "sensors" to monitor the interactions among the components and agents of the product line system and how they influence one another. Ideally, we would like to identify behavioral patterns as they arise.

We need intelligent analytical tools to detect and report changes in system behavior. We require service providers to develop the necessary system sensors. The goal is to identify system behaviors that we may not anticipate. It's about creating an all-encompassing machine learning system to uncover correlations among the components and agents of a product line system. What factors should we consider? Will it include time, money, buying behavior, skills demand, consumer perceived quality, or shifts in the ecosystem? This task will not be easy, and due to the complexity of the system, it will never be executed flawlessly. However, its value could be immense.

PLSM practices must align with the framework and remain consistent with applicable theories. The practices that a product line team aims to enhance should be reflected in the system assemblage mapping. Notice the shift in thinking. The traditional approach focused on expediting the development of individual products. The new approach emphasizes improving the system that creates and manages all products. It's about creating and nurturing the goose that lays the golden eggs, not demanding faster production.

PLSM Praxis and The Accelerating Change Race

As companies advance with PLSM, the next significant contribution will be for the product innovation and management community to expand and enhance the theories and practices, or praxis, that underpin and shape the framework. This effort must engage academics, practitioners, consultants, trainers, and software providers.

Addressing both efficiency and change is essential for most large organizations. PLSM offers an appealing path forward. However, we cannot depend solely on what I have presented in this book. Numerous methods and practices must be developed, and a deeper understanding of the framework's relationship with other topics is vital. This effort requires many individuals to step up. It cannot and will not be managed by a single writer, consultant, or even one academic institution.

Chapter 30 Summary

PLSM Praxis: What's Needed, What's Next

- **Shift focus:** Companies must shift their focus from single products to product lines as complex systems.

- **Stop "Best Practices" Without Strong Theory:** Build the PLSM Praxis that melds solid theory with superb practices.

- **A Call to Action:** The entire product innovation and management community, including academics, consultants, and software vendors, must contribute to PLSM Praxis.

- **New Tools and Methods:** PLSM needs system-enabling tools. These include methods to track how different parts of the system work together and spot new patterns.

- **Building PLSM:** PLSM aims to overcome competition in the Accelerating Change Race, which all major companies must join. It can only do this with robust theory and practices.

Chapter 31

PLSM Resources

Learn, Build, Use, and Improve

Most managers in large companies who develop and manage products are familiar with single-product approaches. The same is true for academics and consultants involved in product innovation and management. I can confidently say that, for the most part, these individuals do not consider multi-product or complex product line systems. However, their foundational knowledge of single-product and structured systems thinking is extremely beneficial for transitioning into PLSM.

A strong PLSM foundation can only take root when managers are motivated to consider product innovation and management from a new perspective. Meaningful exploration can happen only when company managers possess deep knowledge about their products and are inspired to view them in a new context.

This book serves as a primer on PLSM that you can share freely to help managers grasp the topic. However, most organizations need more in-depth information and interactive learning.

Resources

I offer several resources that will open doors to tools, learning, and much more.

- **https://plsm-resources.adept-plm.com**

This is my company's website, dedicated to PLSM. Here, you'll find various services, introductory training, comprehensive master classes, assessments, templates, workshops, and coaching related to the what, why, and how of PLSM.

- **https://www.pdma.org**

The Product Development and Management Association is a well-established organization that offers access to research and focuses on experience-driven single-product innovation and development. The organization's academic foundation helps practitioners understand single-product thinking. However, for the next several years, I would not expect PDMA to provide insights into multi-product complex systems thinking.

- **https://thecynefin.co/**

The Cynefin Company is where Dave Snowden's work on complexity theory and complex systems resides. While Snowden and his company approach innovation and strategy from the perspective of addressing complex problems, they do not claim to be experts in product lines. Snowden's writings, blog posts, and videos effectively help people understand complexity theory and complex systems, regardless of the context.

- **https://strategyn.com**

Strategyn is a training, consulting, and service firm founded and led by Tony Ulwick. They are experts in the Jobs-to-be-Done theory for single products and in defining Verb-based market segments. The concept of Verb-based segments and product innovation charters in PLSM is my intentional adaptation of Strategyn's work.

Online Support Material

Here are some of the supporting documents you may find at https://plsm-resources.adept-plm.com.

Blogs, Whitepapers, and Discussions

Product Line Roadmapping	Practices to Support Non-agent Key Part Roadmapping
System Assemblage Mapping	Practices to Support Agent Key Roadmapping
System Constraint Topography	Getting a Lay of The Land: Plotting Current Constraints
Theory Proof: Product Line System Archetypes & Disruptive Innovation	Theory Supporting Arguments and Counterfactuals
Changing Archetypes	Examples: Impacts and Dynamics
Software Archetypes	A Non-Definitive Exploration of Software Archetypes
Change Capacity	Influencing a Product Line System's Capacity to Change
Flows & Flow States	Pursuing Flow Alignment and Resonance
Management & Governance	The Ongoing Management of Product Lines as Complex, Multi-Product Systems
Chain-link Strategies	The Ongoing Management of Product Lines as Complex, Multi-Product Systems
Key Parts Considerations	The Internal Ecosystem:

Setting Up Sensemaking Affordance	Possibilities and Considerations: What you need to know
Updates on The Accelerating Change Race	Updates on AI and Other Technologies and Market Changes: The Impact on Product Line Outcome Velocity

Advanced Discussion on PLSM

Acquisition Roll Up Strategies	Traditional acquisition roll up strategies seek benefit by eliminating overhead redundancies. With PLSM, we also seek to merge product lines and, where possible, gain greater leverage and better address Verb-based market segments.
Prompts and Agentic Flows	PLSM can benefit mightily when teams understand the use of multiple compounding prompts using flow guidance and team reasoning.
True Transformation	Business and organizational transformations always fall short without building product line Change Capacity. A business can't transform without changing its product line systems.

Narrative Case Studies

GE Plastics	Jack Welch's Ascension: The Impact of System Archetype Changes and Corresponding Business Model Innovations
Keurig Coffee	Hubris as a Force: Human Behavior and Irrational Strategic Moves
Revisiting Kodak and Polaroid	Change Theory: A Product Line Systems View of What Happened
Lennar Homes	Fail: Exploring Platform-Lever Type Changes without System Archetype and Business Model Changes

436

Glossary of Terms

All frameworks have a unique lexicon. If you are an expert in product innovation and management or corporate innovation, terms rooted in complexity science may be unfamiliar to you, and vice versa. You'll also notice some terms that are necessary to describe objects, concepts, actions, and forces at the intersection of complexity and products. This book includes a brief glossary of terms that may be useful to you. A larger, cross-referenced PLSM glossary can be found on my company's website.

APPENDIX

Appendix 1

Artificial Intelligence and Accelerating Change

AI is an awesome technology.

owever, the rate of change is forcing every business, product line, and individual to confront a significant challenge. What's most troubling is that this rate of change is accelerating, and it's likely to increase even more with the advent of Retrieval Augmented Generation (RAG) and Reasoning Agents. These continually evolving Large Language Model add-ons suggest that we're on the brink of AI agents capable of performing many tasks typically done by humans.

We'll likely see artificial superintelligence—intelligence surpassing human genius across nearly every field—enter every company's research and development labs before the decade ends. Claude cofounder and CEO Dario Amodei stated this will happen before the end of 2027 and quickly diffuse into practice within the following year. Furthermore, this remarkable intelligence and its applications will only continue to improve from there.

The influence on product innovation and management will be staggering.

The journey from current conversational AI to superintelligence represents one of humanity's most challenging yet impactful endeavors. Achieving success will require significant computational breakthroughs, and we also must navigate serious philosophical and ethical issues. Nonetheless, there is a growing belief that this journey will continue to drive change for every business and product line.

Consider the statements from Mark Zuckerberg of Meta and Marc Benioff of Salesforce.com. Both have indicated that they will not hire software engineers in 2025 because AI agents can perform the same work much faster and at a significantly lower cost. Software should not impede the creation and development of products.

But this isn't just about generating software, media, or knowledge. It's also about analyzing data and information and creating actions that improve a situation. It transforms variable labor costs into capital investments that enhance value and do not depreciate over time.

The mind-boggling factor is that AI is becoming four times more cost-effective each year. This means it is improving at about one million-fold every decade. Many argue that within a few years, its self-generated improvements will push the four times rate even higher. A five times rate translates into a ten-million-fold improvement over a decade. The exponential upswing, where four or five times per year represents a vertical climb, is sure to change almost every company's product offerings.

Historically, Moore's Law for integrated circuits represents the closest parallel to the rapid advancement of AI. Moore's Law suggests that integrated circuits improve by 2X every two years, resulting in a 32X gain over a decade. However, the comparison falls short once you realize that AI's improvement over the same period is 30,000 times greater than Moore's Law!

An evergreen principle in business strategy has proven to be less enduring than once thought. Wright's Law[104], which suggests that the cost of producing a product decreases by roughly 15% as the cumulative product volume doubles, doesn't seem as timeless. This principle served as the foundation of the S-curve, essential to many strategies. However, OpenAI claims their ChatGPT models experienced a 99% reduction in usage cost during 2022 and 2023 and anticipate a similar decrease for 2024 and 2025. This rate far exceeds the typical price decline suggested by Wright's Law, indicating that entirely new business behaviors may be on the horizon.

AI Forces are Enormous.

In early 2025, SoftBank, OpenAI, Oracle, and MGX announced their investment in the Stargate Project. This new company plans to invest $500 billion over four years to build a more extensive AI infrastructure specifically for OpenAI in the United States. The investment aims to create the computational power, hardware, and software resources needed for OpenAI to process large datasets and effectively develop and deploy AI applications.

The Stargate venture aims to establish a backbone for any AI system, particularly for the United States. Although it is a private initiative, it has the support of the U.S. government to reduce regulatory hurdles and purchase services. To provide context, this venture is significantly larger (by an order of magnitude) than any other new initiative. It also exceeds the equivalent funding of the Manhattan Project by over 15 times. Clearly, every major company will be competing for the ASI capabilities that this venture promises.

According to a report by the McKinsey Global Institute, AI has the potential to generate an additional $13 trillion in global economic activity by 2030. That's more than 15% of global GDP [1], which they say could escalate to 40 or 50% by 2035. This indicates that global GDP will increase by over 10% annually starting in 2030. Such growth is incredible.

MIT professor and Nobel Laureate in economics Daron Acemoglu argues that these projections are neither accurate nor realistic. He asserts that AI can only impact white-collar jobs and will not affect blue-collar work. Consequently, he claims that AI contributes only a 5% positive impact on global GDP, rather than the 15% asserted by McKinsey.

Acemoglu is a highly intelligent economist, and his insightful opinions carry significant weight. However, it's essential to acknowledge that everyone involved in product innovation, development, or management at large companies is considered a white-collar worker. AI, even from Acemoglu's perspective, impacts almost anyone reading this book. If your products cater to or serve blue-collar workers, their demand may not be as significantly affected as that of offerings aimed at white-collar consumers. Nevertheless, it will influence the cost, quality, and performance of those blue-collar products, which in turn will affect buying behaviors and your white-collar job.

This is Big, Real Big

Artificial intelligence (AI) will have a profound impact on product innovation and management. It will reshape customer needs, competitive offerings, and the ways we develop and manage products. It changes everything. This is a new ballgame, one that no one has ever played.

The primary challenge in predicting and assessing AI's impact on product innovation and management is that those who study it often apply their existing knowledge of AI to outdated work methods, using AI to expedite and lower the costs of old practices. However, simply doing more of the same at a reduced cost creates a significant issue.

The challenge for large companies, particularly those that are key players in their industries, is that entrepreneurs and smaller competitors, unencumbered by the organizational inertia of larger firms, can utilize AI and low-cost financing as powerful allies in their product and innovation efforts. The scale and financial barriers to market entry that large companies value will soon vanish.

The Emergent Domain Intelligence Stack

Retrieval-Augmented Generation (RAG) enhances the capabilities of large language models (LLMs) within a company's specific context domain. It achieves this by "chunking and embedding" an organization's internal knowledge base, which encompasses all documents, presentations, reports, and even emails. It does not modify the LLM. Every company should strive to create one or more RAGs to align the AI component system with a specific context.

Reasoning and automation agents also contribute to AI's rapid induction of change. While automation is a simple software integration, a reasoning agent is highly valued. Currently, AI reasoning follows established approaches, including deductive, inductive, abductive, analogical, and common-sense reasoning. A summary of this can be found in an article published in Geeks for Geeks.[95.3]

A reasoning agent specifically designed for complex systems could also emulate a human's use of the OODA loop[58.5]—observe, orient, decide, and act. Additionally, this should facilitate iterative testing. Here we would see:

Observe: Sensing (weak and strong signals), system behaviors, and insight-generating

Orient: Context-specific sensemaking (enabled by a RAG and played off other RAGs)

Decide: Choice generation and iterative testing to select the best action scenario.

Act: Either reporting on choices and recommendations or triggering an action based on automation rules.

Loop: Use reasoning from all previous OODA cycles and repeat the steps.

One key to AI reasoning is that the longer the duration (in minutes and hours, not seconds), the more the AI model and its agents can repeat the loop to test potential actions, resulting in higher-quality recommended actions.

Triggers in the "act" step may include actions such as reporting possible scenarios, modifying the key part roadmap or assemblage map for human (product line team) review and consideration, or instructing the generation of new research and insights.

One critical challenge exists in the realm of AI reasoning. The artificial approach currently finds it impossible to deal with paradoxes. When the AI system encounters a paradox, it outputs inconsistent results or follows programmed logic, which can lead to poor outcomes. Consequently, human intervention must be an integral part of the system. The system must present the paradox to people, allowing human knowledge and experience to make decisions and direct subsequent actions.

Like all complex systems, a paradox cannot be predicted or addressed without human experience. Nevertheless, paradox science is evolving and should aid in resolving paradoxes at the human-AI interface. However, I don't expect a resolution that relies solely on AI.

The pyramid diagram in Figure A1.1 depicts a basic Domain Intelligence Stack for a company and its product lines. It incorporates an industry, a Company, and a PLSM RAG into the LLM. This approach aims to enhance adjacent possible knowledge combinations while facilitating thorough examination. weak signal sensemaking.

AI Valuation

In the future, when a company acquires another, it will be important for the acquiring company to understand the value of its RAGs and AI business agents. The cash flow of existing products will be significantly discounted without supporting AI components. Just as acquisitions must account for intangible assets like brands and trade secrets, they must also consider

447

RAGs and business agents that enhance strategic moves on both sides of the efficiency-change dilemma.

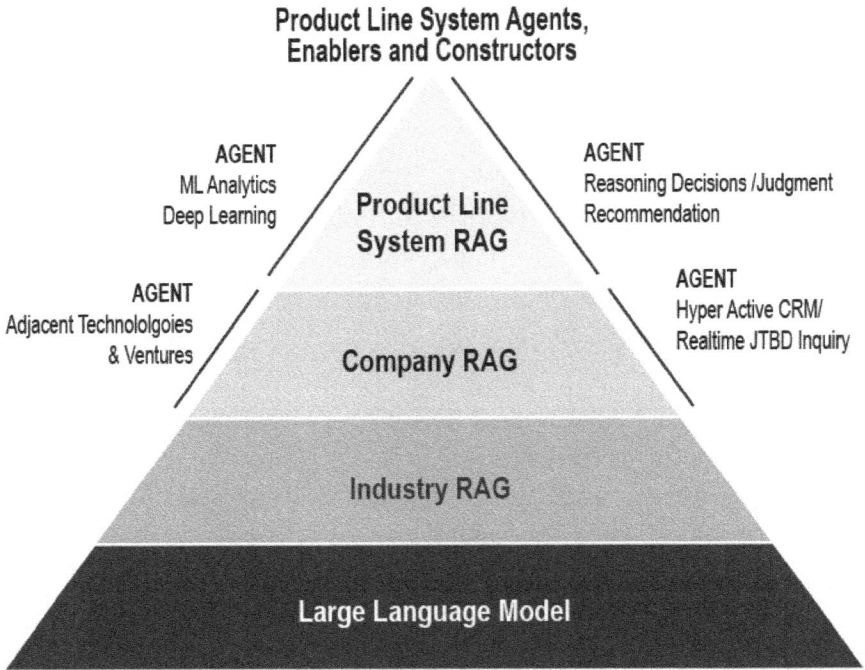

Figure A1.1 Product line systems demand an AI stack specific to the job.
Everyone rents their LLM. The questions big companies face are "How good, in all dimensions, are your industry, company, and product line Retrieval-Augmented Generation modules?" and "How good are the system and AI agents you use to effectuate the Product Line System AI Stack?"

The Battle That Isn't

Taking a step back from product line systems, we can observe a struggle between the forces promoting AI's use and the constraints limiting its adoption. However, the driving forces clearly outweigh the constraints. Tables A1.1 and A1.2 provide an AI-generated view (ChatGPT o1) of these forces and constraints impacting AI diffusion into practice. This is not confined to product line management.

You're entitled to your opinion on AI adoption, but it's at your own risk. If you're right, you'll save some money; if you're wrong, you could go out of business.

Table A1.1 **Driver Arguments Suggesting Greater Change**

Force-Inducing Accelerating Change	Explanation
1. Compression of Research & Development Cycles	<ul><li>AI can rapidly process vast amounts of scientific literature and data</li><li>It can generate and test hypotheses at machine speed</li></ul>This allows companies to iterate on products and solutions much faster than traditional R&D approaches Example: Drug discovery companies using AI to screen millions of molecular combinations in days rather than years
2. Network Effects and Scaling	<ul><li>As AI systems improve, they can be deployed across multiple domains simultaneously</li><li>Improvements in one area can rapidly transfer to others through shared models and architectures</li><li>The more data and use cases AI encounters, the better it becomes, creating a positive feedback loop</li></ul>This creates exponential rather than linear improvements
3. Democratization of Capabilities	<ul><li>AI tools are becoming increasingly accessible to smaller companies and individuals, this lowers barriers to entry and allows more players to innovate</li></ul>

Force-Inducing Accelerating Change	Explanation
	• When more actors can participate in innovation, the pace of change typically increases, For example: Small businesses can now access sophisticated analytics previously only available to large corporations
4. Automation of Routine Tasks	• By handling routine tasks, AI frees up human capital for more creative and strategic work; this allows organizations to focus more resources on innovation and transformation Example: Drug discovery companies using AI to screen millions of molecular combinations in days rather than years
5. Cross-pollination Effects	• AI advances in one industry often have unexpected applications in others; this creates cascading innovations across sectors; For example: Natural language processing advances in tech are spreading to healthcare, legal, and financial services
6. Market Pressure and Competition	• Companies that effectively implement AI can gain significant competitive advantages; This forces other market participants to adapt quickly or risk obsolescence; The result is accelerated adoption and transformation across entire industries
7. Capital Allocation Efficiency	• AI can help identify promising opportunities and optimize resource allocation; This leads to faster market responses and more efficient innovation funding; The result is quicker development and deployment of new technologies
8. Enhanced Decision-Making Speed	• AI systems can process and analyze complex data sets in real-time; this enables faster, more informed decision-making at both tactical and strategic levels • Faster decisions lead to faster organizational changes and market moves

Force-Inducing Accelerating Change	Explanation
9. Reduced Friction in Implementation	• Modern AI tools are increasingly "plug and play"; This reduces the traditional friction in adopting new technologies • Organizations can implement changes more quickly and with less disruption
10. Compounding Effects	• Each of these factors reinforces the others; the interaction between these elements creates multiplicative rather than additive effects; This suggests that change will accelerate non-linearly

Table A1.2 Constraint Arguments Against Change

Constraints Impeding Change	Explanation
1. Institutional and Organizational Inertia	• Large organizations are inherently resistant to rapid change • Legacy systems, processes, and organizational structures slow adoption • Cultural resistance and employee skepticism can significantly delay implementation • Example: Many companies still rely heavily on decades-old technologies like COBOL
2. Real-World Constraints	• Physical processes, manufacturing, and supply chains can't be accelerated indefinitely • Resource limitations and physical laws create hard bounds on certain types of progress • Human biological limitations remain constant despite AI advances

Constraints Impeding Change	Explanation
	• Infrastructure upgrades take time regardless of AI capabilities
3. Regulatory and Legal Friction	• Complex regulatory frameworks require careful compliance AI governance and safety requirements may actually slow deployment • Privacy laws and data protection requirements create implementation barriers • Legal liability concerns can cause organizations to move cautiously
4. Human Adaptation Limits	• Workforce training and reskilling take significant time Human decision-makers often require extensive validation before trusting AI systems • Social acceptance and trust-building processes can't be rushed • Organizations must move at the speed of human adaptation
5. Quality and Safety Requirements	• Critical systems require extensive testing and validation Safety-critical industries cannot sacrifice reliability for speed • The complexity of AI systems may require more thorough testing, not less • Mistakes with AI can be catastrophic, encouraging cautious approaches
6. Economic Constraints	• AI implementation requires significant capital investment Economic cycles and market conditions limit transformation speed • Many organizations lack resources for rapid change ROI uncertainty can slow adoption

452

Constraints Impeding Change	Explanation
7. Technical Debt and Integration Challenges	• Legacy system integration creates significant complexity Technical debt accumulation can slow future changes • Interoperability issues between systems create friction • Security requirements add layers of complexity
8. Data Quality and Availability Issues	• Many organizations lack clean, organized data Building high-quality datasets takes time • Data privacy and sharing restrictions limit AI effectiveness • Historical data may not predict future patterns well
9. Market Saturation and Diminishing Returns	• Early AI adoption may capture most available gains • Subsequent improvements may yield diminishing returns • Market opportunities may become saturated • Competition may stabilize rather than accelerate
10. Complexity Management	• AI systems' increasing complexity may slow development • Integration challenges grow exponentially with system complexity • Managing AI systems requires new organizational capabilities • Debugging and maintaining AI systems become more difficult over time
11. Social and Ethical Considerations	• Public concern about AI impact may slow adoption • Ethical considerations require careful deliberation

Constraints Impeding Change	Explanation
	• Social impact assessments take time • Building trust with stakeholders is a gradual process
12. Natural Innovation Cycles	• Innovation often follows S-curves rather than exponential growth • Historical patterns suggest periods of rapid change followed by consolidation • Fundamental breakthroughs may become harder, not easier • Some problems may be inherently resistant to AI-driven acceleration

We Must Get Onboard

In the 1940s, MIT professor Kurt Lewin developed a method for thinking about change that has endured.[86] He postulated that organizations "unfreeze" and embrace change when the perceived and real gains, plus the perceived losses from not changing, are greater than the perceived and real costs—both psychological and economic—of changing. This is laid out in Chapter 26.

Our current situation with AI indicates significant gains from embracing it and substantial losses if competitors adopt it and you do not. However, every organization is quite concerned about the cost of change and the inertia that slows progress. Still, Lewin's change equilibrium heavily favors change. When your competitors and customers adapt, the pressures pushing you to change increase even further.

The risk of resisting AI in the 'what, how, and why' of our product offerings is, at best, irrelevance and, at worst, bankruptcy.

Appendix 2

Complex and Orderly Systems

A Ferrari is a complicated, orderly system. The Amazon rainforest is a complex system. A manufacturing system is an orderly system that can be either simple or complicated. A product line is a complex system.

Before diving into these two system types, I encourage you not to view this as a binary comparison of one against the other. The most confusing aspect of these system types is that they coexist. Every business is a complex system that can encompass many orderly systems, such as manufacturing lines.

System Types

Understanding system types is fundamental to systems management because each system type dictates the most effective management approach. Each type exhibits distinctly different behaviors. Applying management practices designed for one type of system to another will lead to poor outcomes and results.

Dave Snowden and his Cynefin Company identify five types of systems: random (disorder), chaotic, complex, complicated, and simple (obvious). Consider these types as definitional categories along a continuum, with each type flowing into the next. However, there are no clear boundaries between them.

In product line systems management, we focus on complex systems along with a combination of both complicated and simple systems, which I refer to as orderly systems. I exclude random and chaotic systems because we work with existing systems, and it's unlikely that either random or chaotic systems could survive in a business for very long.

Fractal Nature, Recursiveness, Reductionism

Complex systems are recursive, exhibiting recurring behavior patterns influenced by feedback loops where one aspect depends on another's behavior. They do not display a singular or easily understood linear behavior.

Complex systems exhibit recursive behavior because they are, by definition, fractal. This term refers to a category of geometry in which, when you zoom in on a shape from a distinct point, you observe the same pattern as when you zoom out. The behavioral pattern among components recurs at all levels of magnification. Therefore, the key to managing complex systems is to understand the behavior pattern and identify what influences favorable and unfavorable changes.

The nature of a complex system implies that you cannot examine its components independently to understand the whole. Another way to express this is that complex systems are not "reductive." In contrast, orderly systems are reductive. This distinction poses a challenge in management because most business practices are based on the assumption that systems are orderly and reductive. In many business systems, such as product lines, this assumption is incorrect. This results in practices that don't align with the system.

Management practices must uncover behavior patterns in complex systems and help people understand the interactions among system components. The challenge is that these resulting behaviors often appear in unique and unforeseen ways.

The descriptions below highlight the differences between orderly and complex systems.

Orderly System:

Structure: In an orderly system, the components and interactions are intricate but follow a deterministic cause-and-effect relationship. The relationships between elements are often linear, and there is high predictability.

Fractal Nature: An orderly system is bound by a set geometry. It is not fractal. There is only one non-recurring view of an orderly system.

Behavior: The behavior of an orderly system may be challenging to understand when it's complicated, but it is knowable and can be predicted if we have sufficient information. We can expect the system's response to interventions. Outcomes are generally reproducible.

Examples: Orderly systems often involve machines, technical processes, or systems with many interconnected but predictable parts. Examples include car engines, computer programs, and manufacturing assembly lines.

Complex System:

Structure: A complex system is characterized by non-linear, dynamic relationships between its elements. The system's components exhibit emergent properties, which means outcomes are not guaranteed or predictable.

Fractal Nature: A complex system is NOT bound by set geometry or a single essence. It is fractal, and there are numerous ways to examine complex systems. Each positional view is referred to as a substrate that has a context that differs from other substrates or fractal positions. The context and substrates exist because of a complex system's fractal nature.

Although we often refer to something as a complex system, we are actually describing that system's substrate. If we want to explore the system in ways beyond the discussed substrate, we need to identify an alternate one. For instance, a project management perspective is a substrate within the broader product line. Appendix 3 examines the product line as a substrate within a specific context.

Behavior: The behavior of a complex system is inherently unpredictable and may involve emergent phenomena that are not easily deduced from the properties of individual components. These systems often exhibit adaptability, self-organization, and sensitivity to initial conditions.

Examples: Complex systems exist in both nature and social structures. Examples include ecosystems, the human brain, and market economies. Numerous variables influence these systems, and their behavior is shaped by the interactions of diverse elements in ways that are not entirely deterministic.

Key Differences

Predictability:

Orderly System: With a sufficient understanding, the behavior of an orderly system becomes predictable. Cause-and-effect relationships can be identified, which allows for a high level of predictability.

Complex System: The behavior of a complex system is inherently unpredictable, and outcomes may display emergent properties that are not directly deducible from the characteristics of individual components.

Interconnectedness:

Orderly System: Components within an orderly system are interconnected, yet the relationships are often linear and clearly defined. Changes in one part of the system produce predictable effects on other parts.

Complex System: Interconnectedness in a Complex System involves non-linear relationships. The interactions between components are dynamic, and small changes can lead to disproportionate and unexpected consequences.

Adaptability:

Orderly System: Orderly systems have limited adaptability. Modifications to the system often necessitate a comprehensive understanding of its structure and are generally planned and controlled.

Complex System: Complex systems demonstrate adaptability and self-organization. They can respond to changes in ways that may not be entirely understood or controlled, and new behaviors or patterns may emerge spontaneously.

Reduction:

Orderly System: Orderly systems can be divided into their components, and these components can be optimized and reassembled into a system that performs better. Orderly systems are reducible.

Complex System: Complex systems should not be expected to perform better simply because one part, however you define it, has been improved. When reassembled, a complex system may exhibit different behaviors. New behavior patterns may emerge—Mary Shelley's Frankenstein reassembled a monster, not a man. Complex systems are not reducible or reproducible.

Holistic View:

Orderly System: An orderly system can be viewed as a combination of its constituent parts. A holistic perspective of an orderly system considers its parts, constraints, and the forces that influence it.

Complex System: Complex systems lack a whole. You can always broaden the system to include more elements. In complex systems, we define substrates to establish logical, yet often flexible, boundaries and limits. To comprehend a complex system, you must understand the substrate you are addressing. There is no single, comprehensive view of a complex system.

The distinction between an Orderly System and a Complex System lies in the level of predictability, the nature of interconnectedness, and the system's capacity to adapt to changes. While complicated systems involve intricate yet predictable relationships, Complex Systems exhibit non-linear dynamics, unpredictability, and the emergence of new behaviors.

Hindrances and Improvements

A root analysis method or practice (Lean, Theory of Constraints, Fishbone, etc.) can serve as a viable approach to overcoming obstacles, bottlenecks, and roadblocks in orderly, complex systems. However, in a complex system, no single root cause exists. Everything influences everything else, prompting analysts to observe behavior patterns in the system that either assist or hinder its performance.

The difference in system characteristics makes managing the two system types quite distinct. In orderly systems, you remove or mitigate the constraining cause to enable better flow. In complex systems, you find ways to dampen poor system behavior and encourage better behavior.

Influencing Complex System Dispositions, Propensities, and Tendencies:

For complex systems, we seek to identify behavior patterns within the system that improve outcomes and foster those behaviors. When we identify behavior patterns that hurt the system's outcomes, we seek to eliminate them. I refer to this as "weed and feed" management approach to complex systems.

Complex systems have components and agents that can be disassembled. But if you take them apart, you'll find you can't reassemble them exactly as they were. Each assemblage is unique. By changing the system's assemblage, you also alter the system's behavior, sometimes significantly.

Analysis

All is not lost for managing complex systems. We can analyze complex systems. However, the analysis differs from the root cause approach used for orderly systems. Instead, we focus on the interactions among the system's parts and agents, as well as the behavior patterns exhibited by the system.

It's interesting to explore how the assemblage of a complex system may differ in the future. However, as a general rule, future exploration is less valuable than examining how the system might be influenced toward a novel assembly today. Please don't throw your hands up in frustration when dealing with complex systems. Whether you understand the topic or not, we've learned it is a science. Therefore, you shouldn't give up on the complex view and revert to orderly systems thinking.

Complexity Theory and Product Line Systems

A fundamental principle in complexity theory significantly impacts how we manage product line systems. Each substrate establishes a unique context for its actors to operate within. Most importantly, this context is shaped by the constraints and resulting forces that affect the substrate.

To survive, the system must facilitate viable (and valuable) outcome emergence while managing constraints and potential negative forces. This means that the system's structure or architecture must evolve alongside changes in its constraints. In product management language, this principle is like saying: If your product line system sucks, it will go bust, and the system will die. For a product line system to survive, it must change in harmony with the shifts occurring in constraints and resulting forces. To understand what you should do with a product line, identify the changes occurring in the constraints and the forces affecting the line.

Product Lines

In complex systems, including product lines, forces arise from internal or external factors, mechanisms, or influences that shape, drive, or hinder the emergence of outcomes. Complexity theory experts refer to these sources as constructors and constraints. Positive and negative forces influence the emergence of outcomes, resulting in improved Outcome Velocity for a product line. However, the product lines share certain forces that differ from those affecting other complex systems.

These forces differ from those that affect environmental systems, like the Amazon rainforest and climate change, or broader social systems such as healthcare and education. This distinction reflects the extensive scientific exploration of complex adaptive systems in contrast to the engineering applications of that science within the more focused contexts of multi-product innovation and management. The specific application of complex systems to product lines must be logical and consistent with our understanding from scientific or theoretical perspectives.

Influencing Factors

Forces and constraints interacting with the system's components and agents impact the outcome emergence of a product line both partially (one-off or temporary) and entirely (the behavioral pattern).

1. Constraints: A product line's constraints define it as a system. If constraints change, the system will change.

2. Environmental Factors: The Product Line Substrate encompasses external conditions—market demand, competition, regulations, state disturbances, customer behavior, trends—along with all internal conditions, including flow hindrances, resources, constraints, people behaviors, and bottlenecks.

3. Agent Interactions and Relationships: The nature and strength of the interactions between the system's components (agents), as well as their relationships and connections, determine the system's behavior. If you want to change the system's outcome, you'll need to alter the interaction of the system's agents.

4. Feedback Mechanisms: Product line-related positive and negative feedback loops can act as driving forces, either amplifying or dampening certain behaviors or trends within the system. Think of a KPI as a dashboard reading of a feedback loop. Outcome Velocity is a three-factor vector that represents the product line's performance. Outcome Velocity serves as the primary feedback mechanism for a product line.

5. Initial Structure and Conditions: The system architecture can exert forces that shape the subsequent outcomes and Outcome Velocity direction. The architecture starts with and extends from the product line's archetype.

6. Surrounding Structures and Conditions: The system's structure, the interconnections of its components, and the ecosystem in which it operates can impose forces that either constrain or facilitate the product line's outcome emergence.

7. Agent-level Behaviors: The system agent's behaviors and relationships collectively generate forces that influence system outcomes and performance.

8. Complex System Forces: Complex product line systems forces are internal or external factors, mechanisms, or influences that can shape, drive, or constrain the emergence of specific patterns, behaviors, or outcomes within the system.

Here are some sources for a deeper dive into complex systems:

- **Complexity: A Guided Tour:** Mitchell, M. (2009); Oxford University Press.

- **Business Dynamics: Systems Thinking and Modeling for a Complex World**: Sterman, J. D. (2000); Irwin/McGraw-Hill

- **Systems and the Learning Sciences** (Chapter 26): This chapter reviews the multidisciplinary study of complex systems in physics, biology, and social sciences. It reviews three topics: first, research on how people learn how to think about complex systems; second, how learning environments themselves can be analyzed as complex systems; and finally, how the analytic methods of complexity science, such as computer modeling, can be applied to the learning sciences.

- **Complex Systems and the Learning Sciences.** This downloadable book addresses complex systems and their relation ot learning. You can find it at https://www.cambridge.org/core/books/cambridge-handbook-of-the-learning-sciences/complex-systems-and-the-learning-sciences/C8BAE6593CEE84A57913227796C8530B

- **Complex Systems: A Survey** This is a survey overview of complex systems and their behavior. It includes classic examples such as condensed matter systems, ecosystems, the economy and financial markets, the brain, the immune system, granular materials, road traffic, insect colonies, flocking or schooling behavior in birds, and more.

- **Complex Systems Science and Social Work | Encyclopedia of Social Work**: This article introduces a formal approach to systems thinking, provides an overview of central concepts in complex systems analysis, and concludes with an in-depth example of an agent-based simulation model, which puts complex systems thinking into action in a research and practice context.

- **Complexity Explorer:** This is an online educational resource that provides free courses and tutorials on complex systems science. It is run by the Santa Fe Institute, a research center dedicated to studying complex systems. 6. [Complexity Digest]: Complexity Digest is a weekly newsletter that updates the latest research and news in complex systems science. It covers many topics, including physics, biology, economics, social sciences, and more.

- **Journal of Complex Systems:** This journal is a peer-reviewed academic journal that publishes research on complex systems science. It covers many topics, including physics, biology, economics, social sciences, and more.

- **Santa Fe Institute:** The Santa Fe Institute is a research center dedicated to studying complex systems. It brings together researchers from various disciplines to collaborate on research projects and develop new approaches to understanding complex systems.

- **Complexity Explorer YouTube Channel:** This YouTube channel provides free video lectures and tutorials on complex systems science. It covers a wide range of topics, including physics, biology, economics, social sciences, and more.

- **The Complex Systems Society:** This is an international scientific society that promotes the study of complex systems. It organizes conferences, workshops, and other events to bring together researchers from a wide range of disciplines to collaborate on research projects and develop new approaches to understanding complex systems.

- **Complexity Science Hub Vienna:** This is a hub or center dedicated to researching complexity science. It brings together researchers from various disciplines to collaborate on research projects and develop new approaches to understanding complex systems.

Appendix 3

Complex System Substrates

A system's "substrate" refers to the basis on which the system operates and evolves. For example, a physical-biological substrate would be based on cellular structures and large organic molecules (such as peptides, enzymes, and DNA), displaying unique and sometimes unpredictable behaviors.

Complexity experts identify four basic categories of substrates: physical, environmental, technical, and conceptual. A product line serves as a conceptual substrate that belongs to the technical category (referring to technologies used to). A system's "substrate" refers to the basis on which the system operates and evolves. For instance, a physical-biological substrate would depend on cellular structures and large organic molecules (such as peptides, enzymes, and DNA), displaying unique and sometimes uncertain behaviors.

Product Line Substrate

The commonality of product line constraints and forces allows us to treat product lines as a substrate type, enabling us to build and apply a shared understanding. However, this does not imply that understanding one product line system guarantees a comprehensive understanding of every product line. Instead, it suggests that a common framework of principles can be applied to manage all product line systems.

However, each product line system substrate creates a unique context in which its human actors must operate, and the agents and key parts of the system interact. Most importantly, the details of its constraints and forces define the context for managing it. This same foundation—namely, the constraints and forces—indicates that the system substrate and the context for thinking about it are closely linked.

The management context is a more finely tuned version of the substrate defined by the details and interactions (behaviors) of the system's constraints, forces, key parts, and agents.

Subsystem substrates can either intersect with or reside within a complex system substrate. Interestingly, this substrate may belong to a different category, such as orderly-complicated or chaotic.

The trap many managers fall into is that while the substrate is complex, it can and often does contain subsystems that are more orderly than complex. An issue arises when managers extrapolate the orderly system to represent the entire substrate. These managers may become ensnared in their ordered system perspective and overlook potential opportunities presented by product lines through multi-product complex systems management.

Communicating Substrates

Those managing a product line system substrate must share a "near-common" understanding of the substrate. The same holds true when discussing system opportunities and challenges with others. If one person refers to a substrate or a sub-system within a substrate that differs from another's perspective, the discussion will become nonsensical. My experience shows that this often occurs among experts, who talk past each other without a mutual agreement on the substrate.

In managing product line systems, the various and frequently conflicting substrates include:

- Single-Product Management
- Project Management: Agile and Waterfall/Phased
- Project Portfolio Management
- Market Management
- Manufacturing, Operations, and Production Lines
- Technology and Intellectual Property
- Functional Management and Strategies – Sales, Finance, Engineering, IT, and Human Resources
- Supply Chains
- Market Research, VOC, JTBD
- Partnerships and Ventures
- Business Unit Management and Strategy
- Corporate Management and Strategy
- Leadership
- Corporate Innovation

These substrates pertain to product line systems, with some being loosely coupled and others tightly integrated. What is deemed "best" for each substrate often conflicts with what is optimal for another.

Viewing project management, innovation, or problem-solving through the lens of a complex system differs from assessing an entire business or its product lines from a complex system viewpoint.

Chapter 2 of this book argues that product line substrates surpass or equal the importance of any others in the pursuit of what is best for a business.

The System Substrate and Management Context

The challenge in product line management is to determine how to respond to the line's constraints and forces. Some may exert a negative influence while others have a positive impact. This requires teams to cultivate a keen awareness and a comprehensive understanding of each constraint and force, as well as their interactions with the system's key parts and agents.

The challenge for complexity science and systems experts is their lack of expertise in the wide range of potential substrate topics. For instance, you might hear a complexity expert discussing biological systems, like a specific type of cancer, while using common terminology from that field. It sounds impressive to someone unfamiliar with complexity science or the specifics of that type of cancer. However, it might seem trivial or silly to the cancer expert.

The opposite is also true. Complexity experts will roll their eyes as they listen to a subject matter expert discuss an orderly system analysis of a complex system substrate.

Complexity Science and Old-School Topics

Complexity science has been with us for only three or four decades, but its utility has notably increased in the last two. However, its diffusion into practice appears to have stumbled over issues related to which substrates to address. This is no small matter.

I suspect that topic experts who have spent thirty or forty years honing their expertise are hesitant to engage with complex systems thinking or its foundational complexity science. Doing so changes the thought context in which they are viewed as experts. It's amusing to see professionals in organizational transformations, innovation, and strategy—all fields aimed at initiating new approaches—withdraw from discussions when complexity science is introduced.

It's equally amusing to watch complexity science experts approach a topic with their problem-solving methods, despite lacking deep knowledge of the subject. It's like watching a mathematician design and construct a building. When completed, the mathematician will glow with accomplishment. However, the architect and engineer may deem the approach as wasteful and far from optimal.

Now that complexity science has evolved and matured sufficiently, the task is to integrate complexity into traditional disciplines. In this effort, the initial step is to ensure you're working on the best substrate.

My biggest frustration is that neither product and innovation experts nor complexity theory experts widely embrace the multi-product, complex product line substrate. Nonetheless, I argue it is the most important substrate related to innovation, product management, and business performance.

Appendix 4

Futurecasting The Product Line's Outcome Velocity

Scenario Narratives and Uncertainties

It is impossible to predict the outcome of a complex system with perfect accuracy. However, one can gain a sense of both the direction of its outcomes and the speed of change in those outcomes.

This observation encourages us to differentiate between forecasting an orderly system and futurecasting a complex system. The former relies heavily on quantitative data, while the latter focuses significantly on qualitative aspects.

Futurecasting is a team's best speculation about outcomes concerning internal and external influences and scenarios. The futurecast presents the most likely scenario and its outcomes, accompanied by a narrative explaining why it is considered the most likely, as well as the nuanced relationships between the scenarios.

Futurecasting is not about building consensus. Instead, it focuses on identifying differing viewpoints, capturing the variations in scenarios, reshaping and selecting those scenarios, and then articulating the distinctions within a narrative. While communicating numerical values related to Outcome Velocity is important, it should only be done as a way of communicating within the context of likely scenarios.

The product line Outcome Velocity is a team's self-created, future-looking performance indicator. It's only as good as the thoughts and insights a team brings. In practice, it's an exploration of known-knowns and known-unknowns along with a probing recognition of unknown-unknowns.

Three-Dimensional Vector

Product line Outcome Velocity is a vector that has both magnitude and direction. It reflects changes related to the overall product line's cash flow, customer satisfaction, and competitiveness. All three aspects are crucial. Our challenge is to assess the three coordinates: their positions in the past, their current state, and where the team anticipates future values will lie, based on strategic actions and potential influences.

The three dimensions may auto-correlate to some extent. Poor and good values in one dimension can influence the others unless they are deliberately managed not to do so. Consider Amazon. Financial analysts labeled the company as a 'not-for-profit' for many years while its customer satisfaction and competitiveness surged. Eventually, the company opened the cash flow spigot and immediately invested the money flow into new ventures.

Some managers may view evaluating a product line's Outcome Velocity as more conceptual than practical. However, that isn't the case. Gathering and organizing information that influences scenarios affecting the Outcome Velocity vector is incredibly valuable. The goal is to communicate a judgment regarding the Outcome Velocity along with the reasons, methods, and nuances behind that judgment.

The approach starts as an empirical exercise, aiming to compile insights and confidence levels. It is essential to identify both disagreements and agreements regarding scenarios in this exercise.

The Math Part

The Product Line Outcome Velocity is a vector that reflects changes in a product line's performance across three dimensions: cash flow (x), customer satisfaction (y), and competitiveness (z). Each dimension has both a starting value and an ending value. These values are scalar (single-point values). The scalar values for each dimension can be plotted on one axis of a three-dimensional graph. The product line Outcome Velocity vector is represented by the line or arrow connecting the starting time point to the ending time point.

$$(X_{t0}, Y_{t0}, Z_{t0}) \longrightarrow (X_{t1}, Y_{t1}, Z_{t1}) \text{ t1-t0 is time period duration}$$

Where   X = Cash flow

Y = Customer Satisfaction

Z = Competitiveness

Figure A4.1 *Product Line Outcome Velocity Vector Calculation*

While we present the values as numbers, much of the calculation remains subjective. We utilize these numbers to foster deeper discussions, explore assumptions further, and build a shared understanding of a product line's performance and change.

You can choose any time duration you like. I suggest considering a short, medium, and long-term view, such as 18 months, 30 months, and 60 months.

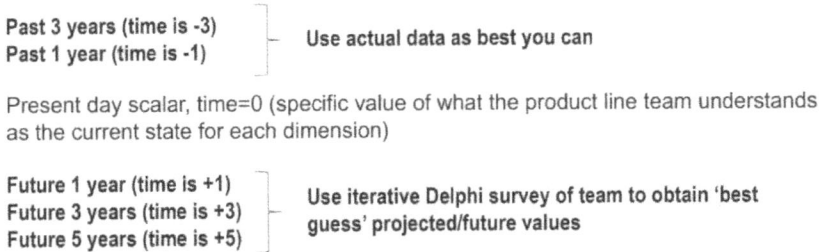

Past 3 years (time is -3)
Past 1 year (time is -1) — Use actual data as best you can

Present day scalar, time=0 (specific value of what the product line team understands as the current state for each dimension)

Future 1 year (time is +1)
Future 3 years (time is +3) — Use iterative Delphi survey of team to obtain 'best guess' projected/future values
Future 5 years (time is +5)

Current Objects	Cash Flow	Customer Satisfaction	Competitiveness
Σ PICs			
Σ PIDs			
Σ PIMs			
Synergy Boost			

Figure A4.2 *Vector Value Estimations For Past, Current, and Future Casts.*

A team of two or three people should gather and compile these quantitative values. I refer to this as the Outcome Velocity team. Team members may include individuals responsible for your organization's portfolio management or roadmapping group.

At the same time, data is being compiled, and the issues and opportunities are captured in a Delphi-like survey (https://www.delphi.com/). Please note that iterative Delphi surveys usually aim for consensus in decision-making. However, this is not how we apply the iterative survey. The team conducts this to capture various perspectives and awareness of the issues and opportunities.

After two rounds of reviews, team members compile and analyze the key factors affecting Outcome Velocity across three distinct time frames: near-term, mid-term, and long-term.

Once the "best guess" data, concerns, and considerations are compiled, the team utilizes the above math templates to calculate the Outcome Velocity for each time frame.

The Qualitative Narrative

An Outcome Velocity is meaningless without a narrative that explains the story of knowns and unknowns, considerations and concerns, and facts and uncertainties. There is no template for creating it, nor a standard format for sharing it; the Outcome Velocity narrative and its presentation are up to the team.

The common narrative guideline for creating a narrative is to convey what the team considered when calculating the Outcome Velocity. The key is to do this as succinctly as possible. Narratives are crucial because understanding a product line system's Outcome Velocity requires knowing what the team took into account to calculate it.

Perhaps the most important aspect of the narrative is that it explores the key components of the line's roadmap, assemblage map, and constraint map in relation to each other and the team's accumulated insights. There is much to this.

Presenting the Outcome Velocity

The presentation of the Outcome Velocity and narrative should also be iterative, involving those on the product teams and key players across the ecosystem (internal chain-links and external companies). However, the team must first address issues related to "need to know," trust, and transparency. This is important because the iterative process may not be pretty; it's more akin to sausage making. Releasing incomplete ideas can be harmful, leading others to mistakenly treat them as final statements.

When the Outcome Velocity is presented, there's a strong likelihood that people outside the Outcome Velocity team will dismiss it as nonsense. Don't discourage such reactions. Instead, use the response to engage the person in explaining what's wrong or what hasn't been considered. Then, carefully break down their comments and how they might affect the Outcome Velocity and related considerations.

The Outcome Velocity evaluation cycle emphasizes managing the line based on knowledge and principles instead of seeking consensus. Ultimately, it falls to the product line team to use their understanding of futurecast outcome velocities to improve the line's performance. At some point, the team must make choices and decisions regarding the Outcome Velocity.

Appendix 5

Product Line Change Capacity

C hange Capacity refers to the degree to which a product line system can endure changes before experiencing negative consequences. When an unexpected change occurs, this is viewed as system resiliency—the system's ability to adapt when a force or action takes effect. Conversely, when the change is anticipated, it reflects good planning.

In product line systems, Change Capacity is a characteristic of the entire system, including its key parts, agents, architecture, and manager-actors. The system's Change Capacity depends on the knowledge, skills, and abilities of the people managing it, as well as the system's tangible and intangible key parts and agents facing known and unknown forces and constraints.

Resilience and Well-being

Change can manifest as quick-response resiliency. This rapid systemic adaptation is geared toward survival. After a crisis like COVID-19, 9/11, or a catastrophic event, one can expect to hear CEOs declare how proud they are of their employees' resilience. In Chapter 13, I address rapid resiliency for product lines.

Enduring change may follow a resilient response to sudden alterations in circumstances; however, it is primarily driven by leadership and is both intentionally planned and meticulously executed. Its goal is to create or enable the future well-being of the system, signifying a purposeful transition into a long-term, enduring state.

A product line's Change Capacity encompasses both resilience and intentional transitions.

The challenge for large companies with established product lines is that the Accelerating Change Race introduces increasing levels of change to every product line. It requires every product line system to possess an exceptional capacity to change.

Change of the System

A key for large companies is that many managers aim to induce change by implementing significant initiatives that often involve the entire organization. However, this approach can be disastrous. A better strategy—central to Change Capacity—is to influence change at the granular level that defines the system (see Chapter 2, 9, and 22). According to the Product Line Change Theory, change efforts must focus on each component of the product line system, including key parts, enablers, constructors, ecosystems, actors, and constraints. This requires a deep understanding of the product line system's architecture.

Teams must strengthen their system's Change Capacity before the need for change arises. This is achieved through awareness, anticipation, and preparation. It requires a sense of urgency prior to significant strategic moves or a resilience-inducing emergency.

Leverage Versus Change

The primary challenge for most lines is finding a balance between asset flexibility and leverage. While leveraging existing assets may lower immediate costs, it can also increase resistance to change. This is essential for designing and modifying platform-levers.

Teams can only tackle the challenge of leverage versus flexibility by managing their product line as a complex, multi-product system. Traditional single-product thinking cannot address this issue. The outdated approach fails to give managers a way to consider changes for all the agents involved in a product line. This is why companies entrenched in traditional product innovation and management will not survive the Accelerating Change Race.

Accounting Misperceptions

During times of change, the concept of sunk cost often clouds decision-making and judgments. We are taught that sunk cost—what has been invested in existing assets and processes—is irrelevant when looking to the future. Yet, these investments still infiltrate most managers' thoughts when they encounter change. Some may view it as a financial burden, while others may perceive it as a psychological cost. Either way, the influence of sunk cost on change must be diminished if teams wish to strengthen a line's Change Capacity.

We face similar challenges with asset depreciation and overhead cost accounting. How people perceive past, present, and future financial issues is important. While many managers may see this as irrelevant when a product line is on the verge of failure, the perspective changes when performance is high. Such challenges can lead to prolonged disputes when no significant issues are on the horizon. Nevertheless, addressing these concerns is essential for building Change Capacity.

The challenge is to instill proper thinking before changes must be made. This requires proactive training and learning. An "urgency before emergency" attitude toward Change Capacity must take hold. It is as integral to the multi-product systems approach as improving the system's Outcome Velocity. At the heart of this shift is the necessity to embrace surviving and winning the accelerating change race as equally important as delivering near-term cash flows.

Building Change Capacity involves addressing negative forces that stem from entrenched practices, outdated processes, and deeply rooted heuristics and assumptions before a change takes place. This process is referred to as "unlearning" and "unsticking" the system agents and influences.

Recognize that the biggest mistake most companies make—besides being single-product and not multi-product oriented—is postponing change issues until change becomes necessary. A key principle of systems management is that teams must address Change Capacity concurrently with Outcome Velocity Improvement. The two go hand in hand.

Strategic Coherence and Change

It is crucial to find a balance between efforts to enhance Change Capacity and those aimed at improving Outcome Velocity. While work focused on increasing velocity drives system changes, Change Capacity enables the desired outcome to unfold. Product lines require both.

Change Capacity and Outcome Velocity are critical characteristics of the product line system. Both must improve for the line to remain competitive in a fast-changing environment. They are essential and must be included in the line's strategy—the primary enabling constraint of the system.

The essence of what is reflected and how it is communicated is specific to the company and product line. For example, product lines burdened by single-product thinking and relying on fixed assets may need to emphasize strengthening Change Capacity much more than driving outcomes velocity.

Fear, Courage, and Change

The fear of failure or losing money can dampen and even hinder the capacity for change. Today's significant difference lies in the magnitude of failure and the amount of money at stake, both of which can be staggering. Consider our current cliffhanger regarding every company in the automotive industry and the traditional chip manufacturer, Intel. Then, take note of major initiatives such as Meta's $50 billion investment in an AI-integrated nuclear power plant and Tesla's billion-dollar DOJO computer, which aims to support the creation of a new autonomous driving platform-lever.

Over the past decade, the dollar amounts discussed for product lines have increased by several zeros, yet most managers' perceptions of money remain unchanged, apart from the zeros. Capital and equity markets have facilitated this significant transformation. The issue is that these added zeros reflect the accelerating pace of change and underscore the likelihood that businesses will face disruptions driven by new-to-the-industry product line archetypes. However, the fear of disruption rarely sets in before it occurs. To me, that's scary.

The financial and capacity barriers that once protected product lines from competition no longer challenge competitors and startups. Large companies can no longer depend on their size for protection.

Any company that has gone bankrupt likely faced issues with its ability to change its product lines. Firms dealing with these problems are easy to identify, as the business press frequently reports on them.

Companies like Volkswagen face significant challenges for not proactively establishing the capacity for change before it was necessary. Now, we see that VW's efforts to reinvent product innovation and management are problematic as the disruption in the EV market has taken hold.

A Continuous Imperative

Building Change Capacity is not a one-time endeavor but an ongoing process. Just as organizations must consistently strive to improve their operations, they must also work to enhance the Change Capacity of their product line.

A key problem for most managers is that measuring Change Capacity is indirect and not straightforward. Instead, it is evaluated through future-cast proxies, such as the speed of adaptation, success in overcoming disruptions, and the ability to remain competitive. These factors are difficult to measure and lack empirical support. Rather, we assess and combine them into an insightful narrative that describes the situation and outlines the best course of action to enhance it.

Many managers quickly emphasize that factors such as company size, workforce skills, assets, and brand reputation are crucial to a product line's ability to adapt. This is true. However, the situation is complex, not merely complicated. What matters is how these factors and others interact to enable or hinder product line system change. This is more a story than a numerical graph.

Assessing Change Capacity

Change Capacity cannot be measured by direct mathematical methods. Instead, it is evaluated qualitatively, which involves a thorough understanding of the situation, as well as the willingness and readiness to act, along with the speed and cost associated with the desired system behavior. This assessment functions as an audit based on various scenarios that impact the system.

The diagram in Figure A5.1 illustrates a workflow for engaging teams and the larger organization in a thorough exploration of their product line system's capacity for change. This process involves building a collective narrative that emphasizes the costs and time required to modify specific system key parts, constructors, and agents. My firm offers customized templates and practices to streamline this workflow. You can find more about this in Chapter 31.

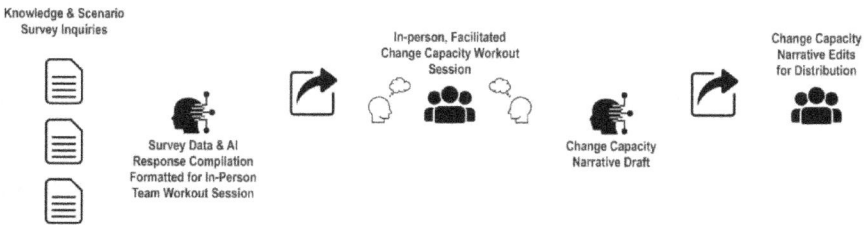

Knowledge & Scenario Survey Inquiries

Survey Data & AI Response Compilation Formatted for In-Person Team Workout Session

In-person, Facilitated Change Capacity Workout Session

Change Capacity Narrative Draft

Change Capacity Narrative Edits for Distribution

Figure A5.1 Workflow (an enabler) for creating the Change Capacity Narrative

This flow engages key players in exploring issues that impact a product line's capacity to change before the change is needed. It combines surveys, AI tools, and workout sessions. My firm offers specific templates and apps to facilitate this flow. You can find more on this in Chapter 30.

Full System Change

Every change, regardless of its magnitude, impacts the entire system. Building Change Capacity requires teams to understand the assemblage and behaviors of their product line systems. They must recognize how different change scenarios might alter the structure. The goal is to identify potential assemblage changes and, where possible, remove or mitigate the obstacles and constraints to these changes. Almost always, the obstacles and constraints lie at the lowest level of granularity of the system's parts and agents.

Without a deep understanding of a product line system and its assembly, businesses must depend on luck and intuition to manage changes. Unfortunately, this is a risky approach that should be avoided.

Scenario Creation and Analysis

Forming scenarios is directly linked to insight generation and sensemaking, as discussed in Chapter 17. The quality of the organization's signal sensing—whether strong or weak—significantly impacts the quality of the scenarios. Evaluating insights based on certainty and uncertainty, as well as related to forces (magnitude and direction) and constraints (hard or permeable), is essential.

Packaging insights into scenarios is an art based in logic. However, teams must feel confident about the breadth and depth of their insights before they start packaging. While individuals play a significant role in this task, valuable scenarios arise from cross-knowledge vetting. This knowledge should extend beyond just the product line team.

Once a scenario is presented, the product line team must identify the necessary changes for the product line to effectively adapt to and capitalize on the defined situation. The team must answer the question: What parts and agents need to change, and in what way, to successfully address the scenario?

The challenge lies in identifying which changes are common across scenarios and which modifications can be implemented without adversely affecting the performance of the current product line. Teams should also assess whether there is a new agent or constructor that would facilitate rapid, low-cost changes that broadly align with the scenarios.

Carrying Out Changes

A crucial part of preparing for systemic change involves reviewing the product line roadmap and the system assemblage map. Each component and agent on these maps may require adjustments— whether minor tweaks or major transformations—to implement the change. Oversight of such change demands governance by executives who understand the system's assemblage and the interconnected nature of its parts and agents.

A key obstacle to change is that many product lines are not designed for evolution. Managers inevitably confront the "change equilibrium," as MIT Professor Kurt Lewin proposed many decades ago.[83] He states that change will occur only when managers perceive the benefits of change to outweigh the economic and psychological costs associated with it. The higher the Change Capacity, the lower the psychological and financial costs of change.

In complex systems, we use a different but complementary approach. Change involves adaptation, and the system tends to adapt by choosing the path that requires the least energy and time. Observe how this relates to the pain incurred by not making the change and the gains that can be realized from it.

Communicating the Change Capacity Narrative

Everyone involved in the change must clearly understand both the pain and the gain associated with it at the time of the change. However, enhancing the product line's Change Capacity reduces the economic and psychological costs of change—the energy and time required to implement change—before it's necessary.

As Lewin pointed out, change revolves significantly around the desire and willingness of others to change. This occurs only through communication.

A Change Capacity narrative encompasses numerous nuances that teams must communicate (and defend) to system actors, key ecosystem players, and the governance team. It conveys the situation and necessary actions to enhance the system's Change Capacity. The workflow intentionally refrains from generating a numerical value as a measure of Change Capacity.

Embrace Change Capacity as a Strategic Asset

Change Capacity is as essential to a product line as the line's Outcome Velocity. Increasing it requires teams to consider all system components, anticipate potential barriers to change, and determine the most effective approach to modifying both tangible and intangible assets. Deliberately enhancing a line's Change Capacity supports its long-term viability and success in driving change. Although intangible, a product line system's Change Capacity is an asset made more valuable by the thoroughness and insight of the narrative.

Appendix 6

Product Line Platform-Levers

"Give me a lever long enough and a fulcrum on which to place it, and I shall move the world." - Archimedes, c. 300 BC

I f you want to sound savvy in a business conversation, simply add the word "platform" after any term. Say "The _____ platform." You can fill in the blank with any term you prefer. It's easy, and it'll sound impressive. This is because the word "platform" implies a powerful force. Unfortunately, not everything that fills the blank is actually a platform. Nor does the title guarantee it will deliver leverage, or that it is specific to a product line.

For our purposes, I replace the term "platform" with "platform-lever." I do this intentionally to ensure you recognize that we are discussing a specific type of component within a product line. Please trust me on this. My experience is that if I do not refer to this product line part as a "platform-lever," the discussions can go off track for some managers and executives.

Platform-levers in Product Lines

Platform-levers play a central role in a product line system's architecture and the line's strategy.

Product lines benefit significantly from leverage because it affects performance in two ways: it accelerates the development of individual products built on the platform-lever, and it improves product attribute performance while decreasing per-unit costs. Fast development combined with improved cost efficiency benefits any product line. No company should deliberately overlook the value of leverage in its product lines.

Shapes, Sizes, and Effectiveness

Platform-levers come in different types, sizes, and shapes. Some provide more leverage than others. You'll also notice that platform-levers have life cycles; they eventually mature and lose their leverage. Unfortunately, a mature, limp platform-lever can greatly burden a product line system.

Customers usually care about products and services, not platform-levers. Yet, these sources of leverage can be incredibly valuable to a company. They are crucial to a company's health.

Table A6.1 below lists ten platform-lever types. I don't mean to suggest these are the only kinds. Over time, smart product line teams have created many new types or classes. Go back fifty or a hundred years, and you'll see production lines as the most common platform-lever. Henry Ford showed us that large-scale production can lower costs and boost performance. Moving forward to today, you'll see many new platform-lever types, variants, and combinations. And those looking to the future will see even more types. The list is not fixed.

Table A6.1: Ten Platform-lever Types

Platform-Lever Type	Platform-Lever Example	Leverage Source
Production Asset	A chemical reactor, a paper-making machine, and just-in-time automation	Scale or flexibility
Hardware Design	A computer motherboard, a system's core controller	Design reuse
Service System	An automated bank teller, an automated car wash, a reservation system	Automation, speed, low cost
Software Systems	A software operating system, integrated application software	Versatility within the intended domain
Proprietary Formula	A unique pharmaceutical, molecular structure, formula, complex system	Uniqueness
Embedded Infrastructure	An optical fiber network, a social network	Connectivity
Modular Systems	A roofing system, a closet connection system, and integrated automation equipment	Adaptability in use, multiple uses
Algorithms, Artificial Intelligence, and Machine Learning	Bank loan screening models, search engines, and voice recognition	Rules, judgment, analysis, decision acceleration, understanding

Platform-Lever Type	Platform-Lever Example	Leverage Source
Connected Integration (IoT)	Intelligent and integrated HVAC (heating, ventilation, and air-conditioning) controls	The data structure, stack, and queues
Blockchain	An NFT that assigns certification to a tangible or intangible product or object.	Scale and meta transfer
Hybrid and Combination	An automotive assembly (with engine and frames as separate platform-levers)	Multiple

Platform-lever Example

To see a platform-lever in action, consider the Samsung "Smart TV." Samsung uses the same motherboard and software, a system-design platform-lever, across different TV sizes and HD performance levels. Their LED screen production platform-lever makes various sizes to increase volume and reduce costs. By combining these two platform-levers (the motherboard and screen production), Samsung built a strong product line system.

The gain sought by Samsung is to develop different TVs more quickly and with improved performance at a lower cost.

Companies stuck in "one-off and repeat" single-product thinking will be vulnerable when competitors introduce new platform-levers. Sadly, most academics and consultants focus only on the "one-off and repeat" approach. Their advice for these situations is to double down by boosting and expanding more "one-off and repeat" development work.

Pivots and Transformations

Because there are many types of platform-levers, strategic moves to change or modify them can often be necessary. And sometimes these changes can be genius. For example, when the strategic move involves keeping the platform-lever type the same but changing the market segments it serves, we call it a product line pivot. A market pivot may not drastically alter the behavior of the product line system, but it can necessitate realignment with different chain-link strategies and business models. I address these dynamics in Chapter 13.

When we change the platform-lever type(s), we refer to the move as a product line transformation. It changes the product line system's archetype—the basic structure of the system. This major change forces companies to rethink and recast their product line strategy and the business model in which the line operates. Accordingly, chain-link strategies must also be realigned. Refer to Chapters 9, 10, 11, and 13 for more information about this major change.

A friend who headed one of the world's largest satellite companies shared with me how he observed such changes in his business. He was particularly impressed with how machine learning was changing his industry.

Here's the backdrop. Traditional large satellites have core designs to which auxiliary technologies add functions specific to customer needs.

In my friend's business, manufacturing each satellite is a highly people-intensive process. They don't have a scaled production because they customize each satellite to a specific and unique end-use. Consider how the military and spy markets demand the latest and most advanced technology for each purchase. This is more of a "job shop" service platform-lever than a design platform-lever.

Orbiting Products

Other satellite markets look different. You'd find that the communication, broadcast, and data handling markets have many needs. Or, as we say in product line strategy, they have many Jobs-to-be-Done. The most common theme for each job is a penchant for reliability or, as the JTBD experts would say, "minimizing downtime." That's where machine learning comes into play. Intelligent analyses of sensor data using machine learning can "predict" issues and failures. This early warning enables the satellite service to take corrective actions before a problem occurs.

However, adding a unique technology like machine learning can present significant challenges. Typically, the first move is to add new technology as a bolt-on to an existing design, regardless of market or industry. It's like adding artificial intelligence to the refrigerator already in your kitchen. The refrigerator wasn't designed for the novel technology, but you do it anyway.

New Technologies and New Platform-levers

The satellite company's job shop and customization approach presents a dilemma. Every machine learning application must be customized to match the unique characteristics of each satellite's data and sensors. Since machine learning works on the principle of creating algorithms from randomized data, there may be some sense to this approach. But how many downtime failures would you need to build validity into the deep learning algorithm?

A more effective approach might be to adopt a core platform-lever design with a machine-learning component that integrates across multiple end applications and customizations.

Here's why the satellite firm needs a new core platform-lever design. The primary Job-to-be-Done outcome customers want is to minimize downtime. How that gets done isn't important. To fulfill this job, the company had to go beyond predictive machine learning, and it's here that they must confront their biggest challenge. The design of each satellite must deliver functional attributes that enable downtime minimization. This means they should reconsider each satellite's design, as well as the boundaries that limit changes to the designs. Confusing, I know. But stick with me.

Jobs-to-be-Done and Platform-levers

The satellite will need the means to automate actions that alleviate anticipated issues. Knowing the impending issue is one thing, but mitigating it is entirely different. Depending on how they perceive the Job-to-Be-Done "downtime minimization," the developers may need to expand the communication system they are working with to encompass the TV in their living room, or perhaps include the equipment needed to communicate with the satellite.

Notice the platform-lever insights shared in the satellite example. First, setting up a platform-lever is a deliberate choice. While design engineers always seek component reuse, elevating a component to become a platform-lever is a conscious, strategic choice.

Leveraging The Jobs

The further a common product element enables multiple products to fulfill sets of Jobs-to-be-Done outcomes, the more leverage it delivers. And the more leverage that's possible from a platform-lever, the more helpful it is to focus on it purposely and intensely.

I don't pretend to be an expert in satellites. However, as a strategist, I wonder if a newly designed platform-lever could be a game-changer. I also wonder how the playing field shifts as competitors purposely create small or mini-satellites with leverage in mind. That's what Space X has done. Is the satellite industry being disrupted?

Disrupting with Platform-levers

Disrupting established industries with new platform-lever types is impactful. That's the essence of the Archetype Disruptor Theory discussed in Chapter 13.

Consider how Tesla, with its multi-billion-dollar battery investment, makes General Motors' engine designs and production infrastructure outdated. Or look at how Apple eliminated Intel to gain greater leverage through its combined M1 chip and iOS platform-leverage. These are significant strategic moves, not just one-off product innovations.

Major strategic moves and pivots, driven by the deployment of new platform-levers, are happening across every industry. Company after company, in sectors like automobiles, medical equipment, consumer goods, services, and software, is upgrading their product lines with new platform-levers. If you have mature leverage or none at all, your company could be at risk. You might find that your competitors, focused on their own pivots or transformations, have you in their sights.

An Intense Focus

For product lines, platform-levers are powerful tools. A lack of sufficient leverage is a common reason why product lines fail. Every company should aim to master this strong performance driver.

It's one thing to say you understand platform-levers. It's an entirely different matter to create and then realize the potential of a superb platform-lever. It is in this aspect of the product line that business genius, or lack thereof, is on display.

The difference between an adequate and a great platform-lever lies in its functionality and how well the organization utilizes it. Creating, developing, and deploying platform-lever can require lots of thinking: creative, strategic, and systems. There is no magic sauce I can offer to create great platform-levers other than to say that it usually takes more intelligence and thoroughness than you first thought.

Using a platform-lever—achieving the highest leverage possible—is just as crucial as the lever itself. Companies must intentionally and vigorously focus on the platform-lever. I strongly recommend establishing a management role responsible for maximizing the platform-lever's potential and ensuring its leverage offers a competitive advantage now and in the future.

The attitudinal mindset required to develop and manage platform-levers extends beyond the single-product innovation approach that most managers learn in school or adopt at work. Major platform-lever initiatives require an intangible element that is never taught—courage. This is because an entire organization might be structured and staffed to pursue a direction that opposes the new platform-lever, and investments in new platform-levers can be considerable.

Platform-lever Genius

The skill and effort needed to incorporate a new platform-lever into an existing product line don't happen by chance. It calls for a review of organizational structures and a rethinking of enabling agents and processes. These important steps only take place when the platform-lever is recognized and managed as a key part of the product line strategy.

You'll see excellence in leading such platform-lever moves that separate the real movers and shakers in a company from the rest.

I encourage all managers and leaders to learn about platform-levers and their role in product line strategies. Please don't expect this knowledge to reach you through trickle-down social media diffusion. Time matters, and you don't want to be the last to adopt platform-levers in your product lines.

Appendix 7

Market Segmentation, Attribute Positioning, & Innovation Charters

Every marketing major has taken a course on product positioning. These classes focus on explaining the benefits of a single product to customers. Unfortunately, they ignore multi-product positioning and do not teach how to create and develop multiple generations of products. They also fail to instruct prospective managers on how to use platform-levers and technologies, scale operations, and reduce costs.

The gap between traditional product management and product line systems management is clear. This difference explains why companies like Tesla and Nvidia have transformed their industries. Once you see this gap, you can't ignore it, no matter the company or industry.

Attribute positioning is where the rubber meets the road in product line systems management. Don't think of this as just a marketing job—it's not. Product line attribute positioning is a deliberate decision that acts as a cornerstone of a product line strategy. All functions, including marketing, engineering, operations, and sales, must implement it.

Product-led Product Line Systems

Being "product-led" is a fundamental principle of product line management. This idea pertains to a situation where the product's appeal, rather than marketing or sales efforts, drives purchases. It guides teams to ensure that each product effectively addresses customer needs. However, in product line systems management, teams must go even further. Every product must also complement and improve the other products in the line.

Unsurprisingly, many teams depend on single-product thinking to develop products (see Chapter 5). While a single new product can be a major success, it is not enough on its own. The single-product approach fails to effectively manage product line systems, regardless of how well these developments align with customer needs.

Instead, teams must deliver a continuous stream of products that improve the line's Outcome Velocity (see Chapter 4). The entire product line system, like other systems, is always more important and valuable than any individual product.

Product line positioning may seem straightforward, such as with basic SaaS offerings divided into free, basic, and professional tiers. However, it gets more complicated in other cases, such as automotive product lines. This complexity involves platform-levers, brands, and technologies. It also encompasses multiple market segments and addresses concerns about shifting customer needs and competitors' actions.

Product Line Positioning–A Different Context

Effective positioning across an entire product line can yield impactful outcomes that are unattainable with single-product positioning. Discovering how to achieve these benefits is true business genius. Fortunately, this accomplishment isn't merely a matter of luck, nor does it arise from a straightforward analytical method.

The genius requires that multiple thought worlds come together and strengthen one another. It calls for a substantial amount of marketing analytics, design thinking, engineering excellence, operational creativity, and strategic foresight. Different specialists will emphasize the importance of each. In practice, however, genius product line attribute positioning requires teams to leverage all of these thought worlds.

To start this essential task, teams are recommended to conduct Jobs-to-be-Done (JTBD) market research. This approach to understanding customer needs also helps identify the features and attributes a product should have to fulfill those needs. The process involves analyzing and grouping customers based on the "Jobs" or outcomes they aim to achieve.

The challenge is that the clusters might seem mathematically correct, but developing the attributes to match the jobs or outcomes can be difficult. Simply knowing the customer's desired results isn't enough. Teams also need to design the solution, operations must manufacture it, and sales must sell it. Most importantly, the solution must align with the product line's archetype while improving the system's flexibility. Refer to Chapters 11 and 22 and Appendices 5 and 6.

Verb-based Segments

Many product managers utilize conventional market segmentation, concentrating on categories such as geography, product types, industries, or price points. These segments are referred to as "Noun-based." In the product line stack (see Chapter 9), you'll notice how we complement Noun-based segments with "Verb-based" segments.

Verb-based segments differ from traditional noun-based market segments. They are organized around Jobs-to-be-Done outcome clusters.[1,2] These are not isolated clusters, as seen in a single-product JTBD analysis; instead, they comprise sets of clusters. Each cluster is a Verb-based segment within the overall market. I use the term "Verb" to represent the action of getting the job done and to distinguish these segments from traditional segments that are named with words that are "nouns."

It is important to recognize that a Verb-based segment may overlap with several Noun-based segments.

Multiplicity of JTBD Outcome Clusters

A key principle in product line systems management is that teams should design the attributes of the products in the line based on JTBD outcome clusters or, as shown in the product line stack, the targeted Verb-based segments. In product management terminology, such engineered positioning makes the line's strategy "product-led."

Along with engineering product attributes that align with JTBD outcome clusters and fit the system's archetype, product line teams must also develop messaging that effectively communicates the product's attributes.

Each product and product change reflects a strategic move aimed at increasing the line's Outcome Velocity. Usually, this move is designed specifically for a Verb-based segment but is carried out based on behaviors in Noun-based segments such as geography and demographics. Notice how managing the product line system requires consideration of both Noun and Verb-based segment types.

Table A7.1. *This table illustrates the relationship between noun-based and Verb-based segments. Notice the word 'minimize' in front of each outcome. This wording helps standardize the survey questions.*

Noun-based Segments	Verb-based Segments (Job-to-be-Done)	Customer-desired Outcomes across all segments
Tablet Computer	■ Creating Presentation Graphics on the go ■ Viewing Media on the go ■ Reading eBooks anywhere	■ Minimize battery outage ■ Minimize operating system crashes ■ Minimize weight ■ Minimize eye strain ■ Minimize thickness
Portable Computers	■ Gaming ■ Creating Media ■ Manipulating Data ■ Running Major Programs	■ Minimize muffled sound ■ Minimize internet to CPU lags ■ Minimize data slowness ■ Minimize thread overload ■ Minimize screen glare ■ Minimize screen reflection ■ Minimize lost wi-fi signal ■ Minimize touch screen sensitivity ■ Minimize fragility ■ Minimize cost ■ Minimize down time ■ ...n, this may be 100+ statements

A common positioning issue underscores the need for organizational alignment. While a team may successfully define Verb-based segments and deliver specific attribute sets to execute the necessary "Jobs", other functions (links in the strategy chain) may not keep pace. For instance, a sales force might still segment the market by noun-based geography, which corresponds with their strategy. In this case, the sales force's approach may not align optimally with the product line strategy, and vice versa.

To achieve chain-link synergy (see Chapter 14), you must address inconsistencies between Noun and Verb segments. Chain-link functions supporting Noun-based segments, like a sales force divided by regions, should also support Verb-based segments. A mismatch could indicate that these functions need reorganizing to better support both the Noun and Verb segments.

The Product Innovation Charters

From a product line system's perspective, each JTBD outcome cluster contributes to an "enabling constraints" (Chapter 15) known as "Product Innovation Charters." These system agents provide guidelines for product concept generation and engineering design (see Figure A7.1). Enabling constraints help teams deliver products that meet customer needs. This work is essential for achieving customer satisfaction and outperforming competitors, two factors vital to increasing a product line's Outcome Velocity.

Noun-based Market Segment

Verb-based (Jobs-to-be-Done) Segment   **Verb-based (Jobs-to-be-Done) Segment**

Customer-desired Outcomes	Customer-desired Outcomes	Customer-desired Outcomes	Customer-desired Outcomes	Customer-desired Outcomes
#1	#3	#2	#16	#14
#2	#12	#18	#11	#10
#17	#7	#9	#2	#45
#26	#52	#52	#34	#47
#52				

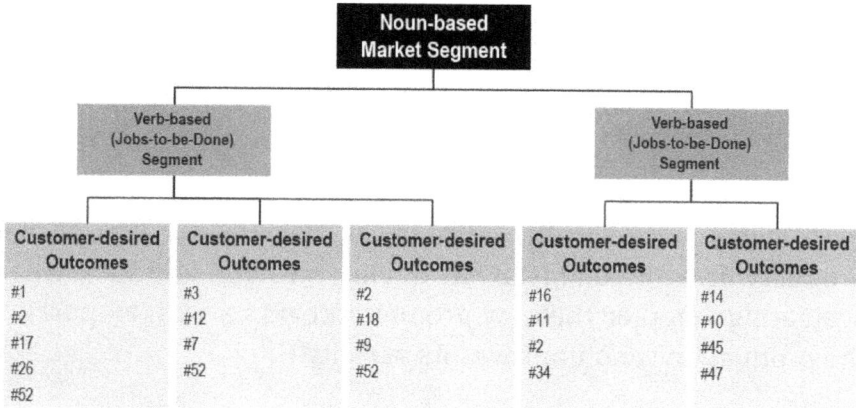

Figure A7.1. *The relationship between segments and customer-desired outcomes is important to a product line's performance. It sets targets for concept generation and product design specifications. Therefore, a team's primary marketing task is to determine clusters of customer-desired outcomes. They do this using quantitative techniques.*

In product line systems management, the line's comprehensive marketing approach cannot rely solely on verb-based segmentation. Other components, agents, and the system's architecture must also be considered alongside various forces and constraints.

Verb-based segmentation alone cannot push a product line past competitors or adapt to specific technological updates. Furthermore, these segments will not promote alignment in work and decision processes or resonate with other business units. Therefore, the charter must improve JTBD outcomes with guiding principles that reflect the product line system's archetype and the business's chain-link strategies.

Each product innovation charter acts as a target or focal point for innovation. It is not just a concept or a product idea. At its core, the innovation charter reflects strategic intent, including both objectives and purpose. Teams should incorporate attribute targets—defined as a JTBD outcome cluster—into each Innovation Charter.

The charter guides development teams in creating a new product that aligns strategically and works seamlessly with other products. It helps teams avoid relying on hope and luck when evaluating whether the new product integrates with all parts of the product line and improves its velocity.

Product innovation charters (PICs) are crucial elements of a product line system, just like products already in the market (PIMs) and products in development (PIDs). PICs define where and how specific offerings intend to advance a product line, making them essential for managing these systems. Without PICs, product line strategies become unpredictable, relying on luck to create the next great product.

The concept of PICs was first introduced in the 1980s by the late Professor C. Merle Crawford, an outstanding innovation pioneer and, I'm proud to say, a good friend. His research demonstrated that having a clear focus on seeking innovation increases the chances of discovering something significant and successfully developing and commercializing it.

Innovation Charter ID QRN-7267
For a new product within the headphone product line

Objective
To gain greater market share in the fast-growing, high-end, long-wear "getting perfect sleep" market segment via noise-canceling + masking wireless earbud

Purpose in Product Line Strategy
Fend off competitors and gain greater leverage from QRZ micro-power platform-lever.

Target Segment(s)
High-end, sleep market—long-wear noise-canceling earbud market segment.

Target Jobs-to-be-Done Segments
Sleeping deeply, comfortably, and free of noise disturbances

Highly Ranked Customer-Desired J-T-B-D Outcomes
Pillow-to-ear comfort level—minimize discomfort
Enduring sleep—minimize sleep physical- and noise-based disturbances
Sweat-free use—minimize sweat and heat buildup
Long wear power without wires—minimize power loss at less than 14-hour use
Ambient noise elimination—minimize < 8,000 Hz seepage

Business Fit Guidelines
Do not diminish any company brand.
Do not position as low cost.
Achieve perception of superior quality compared to alternatives.
Achieve best-in-class noise-canceling parameters.
Achieve buying intent.
Leverage current Supply Chain
Leverage Current Salesforce
Expect New Digital Marketing Capability

Platform-Lever Parameters
Leverage existing QRN micro-power platform-lever with new third-degree material finite modeling for micro battery density and S/N optimization.

Key Roadmap Milestone
Ready for Consumer Electronics Show, Spring 2026

Cost Parameters
Not to exceed fully absorbed costing of $115 per headphone with accessories and case— based on 2nd-year production.

Figure A7.2: Product Innovation Charter—Attribute Positioning
The simple example of an innovation charter on the next page highlights attribute positioning in gray. Most product line strategies demand more information than shown here. Notice how the charter highlights achieving perfect sleep as a Job-to-be-Done, a verb-based market segment.

The focus a team gains through innovation charters eliminates old-school product screening methods. This approach differs from traditional single-product innovation and management. Figure A7.3 shows how the innovation charter connects market segments to product requirements.

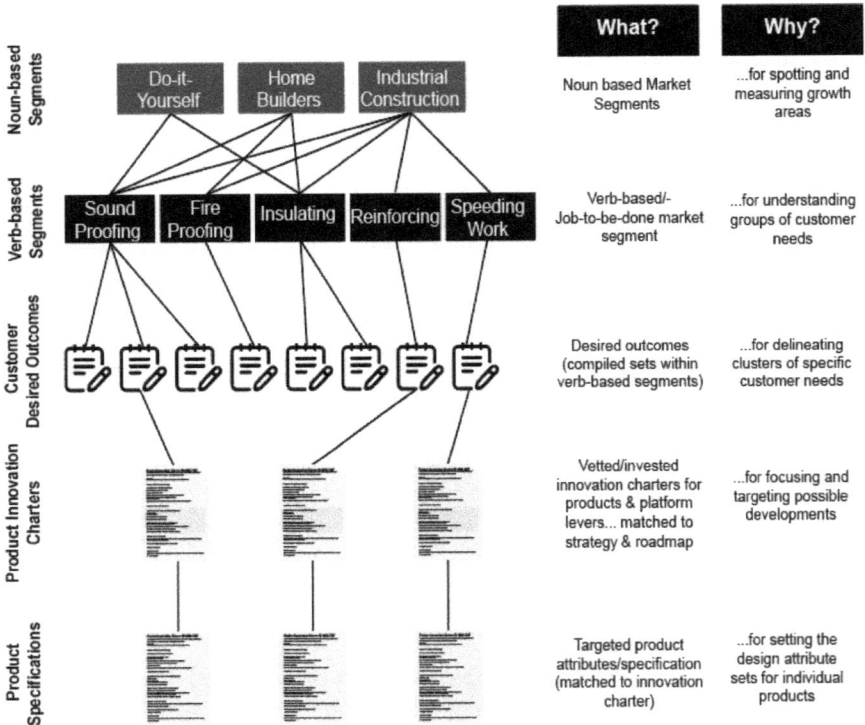

	What?	Why?
Noun-based Segments	Noun based Market Segments	...for spotting and measuring growth areas
Verb-based Segments	Verb-based/-Job-to-be-done market segment	...for understanding groups of customer needs
Customer Desired Outcomes	Desired outcomes (compiled sets within verb-based segments)	...for delineating clusters of specific customer needs
Product Innovation Charters	Vetted/invested innovation charters for products & platform levers... matched to strategy & roadmap	...for focusing and targeting possible developments
Product Specifications	Targeted product attributes/specification (matched to innovation charter)	...for setting the design attribute sets for individual products

Figure A7.3 A Breakdown of the Noun-based to Verb-based Segmentation and Product Attribute Structure. *This example shows the breakdown for a building materials product line. The middle three layers are critical. The example shows customer-desired outcome clusters as Verb-based segments. Innovation charters target these clusters.*

Once innovation charters are added to a product line roadmap, they must also be included in the product line's project portfolio. From this point on, the charter project should be managed for resources and funding, balancing it with all other development projects.

This approach differs from traditional portfolio management practices. With PICs, we expand portfolio management to include innovation charters and no longer manage projects solely within a staged or agile development process. In PLSM, portfolio management deliberately incorporates front-end work.

Innovation Charters and Positioning

A core PLSM principle is to "never jump ahead with a concept." Just because someone comes up with a product idea doesn't mean it should become part of a product line system, regardless of how important the person or people generating it may be. Instead, teams should concentrate on developing innovation charters specifically designed to grow the product line. First, teams "deconstruct" such concepts into innovation charters. Then, they shape the charter using key product line elements and strategy (enabling constraint) guidelines.

This method requires teams to investigate customer-desired outcomes, especially JTBD research covering all segments. The main question is whether the potential innovation charter aligns with a specific Verb-based segment or if the segmentation scheme (how outcomes are grouped) needs adjustment. The success of the team's efforts in this area will affect the quality and compatibility of possible new offerings and products.

A team should promote a charter for further development only when it enhances the speed of the product line's outcome. When a team promotes a charter, they add it to the Product Line Roadmap. It will be linked to a platform-lever, Verb-based segments, and Noun-based segments, along with potential technology building blocks. This indicates that the charter is ready for the investment of time and resources.

Each charter must be connected to at least one Verb-Based segment (JTBD Outcome Cluster) on the roadmap.

For some, the PLSM approach might conflict with everything they've learned in their marketing courses—always prioritize known customer needs. However, innovation charters can target unknown markets just as they focus on unfamiliar technologies. Charters then equip a project team to concentrate on the future and its uncertainties. Most importantly, though, charters should help teams position future products with the right attributes.

Positioning and Branding

Attribute positioning and branding are distinct yet closely related topics. The brand image and message should align with (not conflict with) each product's positioning. However, brand management operates separately, as a chain-link, within the overall business strategy. Don't assume that a brand and a product line are the same.

A product line can include multiple brands, and a single brand can have several product lines. This one-to-many relationship allows teams to develop positioning strategies that are independent of the brand image. For example, consider a branded laundry detergent that comes from a particular product line. That line might serve both the company's own brand and a private-label brand. This strategy can help the line gain greater production efficiency.

Now, let's consider the many dimensions of product line positioning.

Position for Purpose (A Key to Good PICs)

Teams should keep in mind four strategic purposes when positioning their products in Verb-Based segments.

Driving demand. Effective positioning happens when one or more products in a line positively influence the demand for others. For example, consider how the high-priced luxury Mercedes S-550 affects the demand for the lower-priced C-250. While branding impacts the C-250's demand, customer satisfaction with the product is what truly drives that demand. All products should be designed, developed, and positioned to leverage this effect. Platform-levers often play a key role in enabling better cost structures.

Beating competitors. Effective positioning prevents competitors from reaching the critical mass needed to leverage their platform advantages. This strategy directly aligns with fulfilling customer needs through targeted attributes in specific segments. You hinder competitors by selectively satisfying customers. Smart positioning requires teams to consider verb-based segmentation schemes for both customers and competitors. Johnson and Johnson's Vistakon business exemplified this when it launched the first disposable contact lens. J&J expanded its production platform-lever so significantly that it surpassed 75% of the total market. Competitors could not scale enough to compete with J&J's leverage.

Exploiting capabilities. There are many moving targets in product line positioning, and hitting these targets requires more than just setting prices and communicating the benefits of the line's products. Teams must also develop, design, and engineer product features that serve each segment. At the same time, teams must maximize platform-levers and maintain strong chain-link alignment. This approach requires that the team either has (or can develop) the necessary skills in marketing, technology, design, and idea generation to achieve the best results.

Without the right skills, some positioning moves may be impossible. Consider how Dow Chemical used its capabilities when it raced its competitor Exxon Chemical to market with a new polymer (plastic). Dow shifted its focus from using a production reactor as its platform-lever. Instead, Dow built a simulation model to serve as its platform-lever. To make this effective, the company had to change its sales force's focus and align operations with its new approach. You can read this case in Appendix Case Study 3.

Timing dynamics. Beyond today's competitive status quo, forward-looking positioning can also deliver substantial gains. This strategy requires more than just a continuous focus on segmentation and technologies for most product lines. It involves consistently advancing the game plan with a future-oriented perspective. This is where innovation charters come into play. Teams can intentionally establish innovation charters to address evolving or radically new needs. It usually includes multi-generational planning of platform-levers, each linked to expanding market segment positioning.

Review the planned development of Amazon's Kindle product line. The Kindle line includes several platform components: Paperwhite, Fire, Voyage, and Oasis. Each product advances and updates through multiple generations. Amazon designs each generation to meet new user needs, which guide engineering specifications. The timing of each platform-lever generation is key to the product line's success. This strategy must also coordinate with new versions of its operating systems: customized Linux for Paperwhite, Voyage, and Oasis, and Android for Fire.

One-off Positioning Disasters

While all teams should use attribute-based positioning in their strategies, not everyone leverages it fully for their product lines. For example, some managers reactively hold single planning sessions after a competitor's move. This suggests that the team is adjusting its strategy too late, which can be problematic. Moving first and positioning ahead of competitors is much more beneficial. This is called the "first-mover advantage."[5]

Teams need a comprehensive view of the product line to make quick, timely decisions. This strategy covers all potential offerings to each segment, as attribute options evolve and customer groups may shift. For many product lines, this creates a vast and complex landscape. The complexity increases because customers in each segment have different wants, needs, and goals. Competitors constantly improve their products, and internal dynamics can cause each product to head in an undesirable direction. But that's the nature of the game. You can either play to win or face the consequences.

Positioning Tactics for Products and the Line

Figure A7.4 shows the four orientations for positioning products. Remember, this is positioning with product attributes, not messaging. The tactics are:

1. **Price or value.** A team can concentrate its product line strategy on offering an economic benefit to customers by maintaining lower cost structures than competitors. There are several ways to do this. Teams may lower the price of the product or service. They might also create time-based payment options. Alternatively, they can achieve this indirectly over time through the product's or service's "total cost of use."

2. **Quality and delivered performance.** A team can focus on delivering product features that outperform competitors in specific segments. This strategy might also mean that some segments and customers are not targeted.

3. **Speed of benefits/timeliness.** Speed and timing are essential for success in the marketplace. A team might focus on delivering benefits faster than competitors or at specific key moments. Speed affects how quickly customers experience satisfaction from a product or service. Timing involves the release of product features throughout the product or service lifecycle and across different generations of a line.

4. **Offering variety.** Providing customers with a choice (or not) can be crucial to the success of a product line. Different approaches may require different levels of product variety. By increasing variety, more customers across various segments may feel satisfied, and competitors might be kept at bay. Teams evaluate variety within the line by both length (the number of unique products) and width (small variations on products, like size and color).

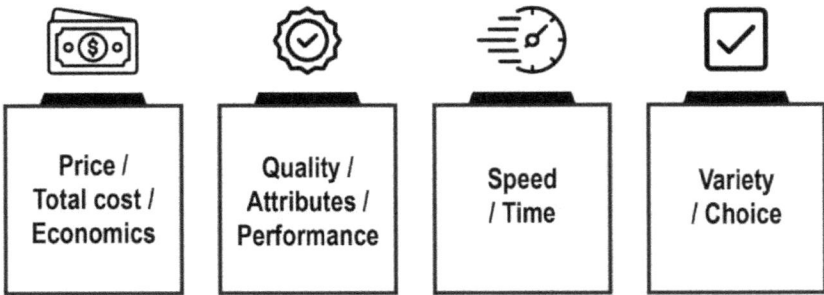

| Price /
Total cost /
Economics | Quality /
Attributes /
Performance | Speed
/ Time | Variety
/ Choice |

Figure A7.4 Product Positioning Tactics *Product line teams can position their products within market segments using one or more positioning tactics.*

Aiming for market leadership with one, two, or three positioning tactics can be a valuable goal. It can increase customer satisfaction and help fend off competitors. However, aiming for dominance in all four factors at the same time can lead to undesirable trade-offs. This all-encompassing approach often falls short because at least one aspect is likely to be weaker. Competitors might take advantage of that weakness by focusing their products on new features that directly challenge the weaker option.

The Exception to the Rule: Blue, Red, and Purple Oceans

In a special case, when organizations introduce a new platform-lever into a market, all four positioning tactics come together. Strategists refer to this as a "blue ocean strategy." This term differentiates it from a "red ocean strategy," where competition exists. We call this a red market because the competition can be fierce.

In a blue ocean strategy, everything is open and new. There is no competition, and product line teams can use all four positioning tactics without triggering a competitive response. Using all four tactics can make it particularly hard for competitors to catch up when they enter the market.

A blue ocean approach often conflicts with an organization's functional chain-link strategies because blue ocean strategies involve a high level of novelty. Blue ocean opportunities are suitable candidates for separation from the organization's existing functional chain-links.

However, blue ocean opportunities are scarce. Most product line teams encounter competition. Many experts suggest that organizations should focus on competing in purple oceans when both advantages and competition exist.

Using attributes to position products is a fundamental principle of PLSM. However, effective attribute positioning also requires alignment with the other two factors: chain-links and the platform-lever. This unity is crucial and influences a product line's speed of success.

Driving The Line's Velocity

Verb-based segmentation is crucial for maximizing product line performance. The same is true for understanding the desired customer outcomes in each segment. Without this insight, positioning can easily go off course. However, knowing this information is of little use if a product line team cannot act on it. What really matters is taking action. This requires teams to invest both time and money and back it up with market analytics.

My experience, along with that of other experts, shows that companies often under-invest in this important part of product line strategies. Instead, they depend on luck and chance to inform their decisions. It is common for companies to invest far less than the potential value that such information and insights could offer.

The problem comes from the single-product mindset that some top managers still adopt. Top management (see side box) is often hesitant to invest large amounts of money or resources if the impact only affects one product. However, in product line systems management, positioning is not about a single product but the entire line, which has much greater value. The research, time, and effort needed for effective positioning are also more expensive than those for individual products.

Building such thinking into the organization's fabric can be a slow process. Some top managers are likely to push back, which should be expected. But this also part of the challenge. That's why a well-planned transition into PLSM, as discussed in Chapter 26, is so important.

References

1. Jobs-to-be-Done theory is laid out by Harvard Business School Professor Clayton Christensen with Karen Dillon, Taddy Hall, and David S. Duncan in their book Competing Against Luck: The Story of Innovation and Customer Choice, HarperCollins, 2016. Christensen and his coauthors also published the topic in the September 2016 issue of Harvard Business in the article: "Know Your Customers' Jobs-to-be-Done."

2. The Jobs-to-be-Done theory builds on "customer-desired outcomes." Tony Ulwick and his consulting firm, Strategyn, developed and promoted the customer-outcome approach. See https://strategyn.com/. Ulwick details this in his book, What Customers Want: Using Outcome-Driven Innovation to Create Breakthrough Products and Services, McGraw-Hill, 2005. Ulwick's recent book, Jobs to be Done: Theory to Practice, Idea Bite Press 2016, details how to benefit from the Jobs-to-be-Done theory.

3. C. Merle Crawford, Professor Emeritus of Marketing at the University of Michigan, co-authored the seminal textbook on new product management with Professor Anthony Di Benedetto of Temple University. Crawford also founded the internationally acclaimed Product Development and Management Association (PDMA), recognizing Professor Crawford by placing his name on its highest academic achievement award.

4. Such customer-desired outcome delineation requires quantitative (not qualitative) research.

5. Michael Spence, a Nobel Laureate and past dean of the Stanford Graduate School of Business, popularized the term "first-mover" advantage in his 1981 article "The learning curve and competition," published in the Bell Journal of Economics.

6. This is the premise of Professor Clayton Christensen's book Competing Against Luck: The Story of Innovation and Customer Choice, HarperCollins 2016.

Narratives

Appendix CASE STUDY 1

Rockwell International

Information and Insight Flow

This case highlights the challenges established companies face in ensuring that key insights flow across organizational functions and hierarchies. It addresses information and insights relevant to product offerings and examines how to convey knowledge to drive new developments. Additionally, it focuses on changes within markets, complicated product lines, and lean practices. The case modifies certain details to maintain client secrecy.

Background

A large business-to-business hard goods manufacturer engaged The Adept Group to analyze how they overlooked an emerging growth market, which we now recognize as hydraulic fracking in oil and gas wells.

The client was accustomed to being number one or two in each market they served. Why, they wondered, were they number six in this new market?

The client employed some smart people. Plus, management had a stellar track record.

Interestingly, many people in the company were aware of the fracking market. However, the company did not develop a product for that market. They aimed to tackle two aspects of their challenge. First, they wanted to understand how and why they failed to respond to the market. Second, they sought to learn how other companies adapted to market changes and capitalized on emerging opportunities.

The Adept Group team conducted a survey of the client organization to gather additional insights. The team also compiled secondary research on the topic and conducted numerous interviews with key managers at other non-competitive companies. Our findings proved to be quite intriguing.

Early in the engagement, it became clear that hydraulic fracking market data was available. A young Market Research manager had compiled a report describing the market. Her report detailed the relationship between the market and two of the company's product lines. Unfortunately, no one or process managed this or any other emerging markets. Our analysis indicated that the information had nowhere to go. The market information was not being transformed into actionable insights to drive product offerings.

What the Research Says

The team's secondary research revealed a topic that had not yet been fully developed. Debriefing the study suggested three summary points:

- First, the ideal scenario is for the person who might use the information to be the same person who discovers it. In such cases, there's no transfer of knowledge. When creating product offerings, this process is best carried out by technical experts conducting market research.

- Second, proximity can play a crucial role in communication and information sharing. Collocation enhances the likelihood of informal 'water cooler' discussions.

- Third, a process flow can help ensure that information moves to the right people in complex settings.

The client's silo structure hindered cross-organizational activities that facilitate knowledge transfer. Furthermore, the separation of offices, despite being on the same campus, limited communication through proximity.

Benchmarking interviews were revealing. We learned that other firms employed effective methods to facilitate the flow of information. However, no firm considers itself exceptional at converting information into meaningful strategic actions. The techniques appear to be effective in a one-off event manner, rather than as a consistent flow. These methods included:

Table CS1.1 Insight Methods and Flow

Competitive Analysis	Customer feedback (general)	Customer Trends - system, product
Economic Analysis	Sales and Partner Feedback	Project feedback (lessons learned)
Geographic Analysis	Technology Trends	Country / Political Risk

Market Access Analysis	Regulatory / Certification / Gov. Trends	Customer Usability Feedback
Customer Commitments	Performance Measures (Shipments. Unit, $)	Customer Loss Feedback

Declaring a cross-organizational process as the solution to the client's challenges was insufficient. The Adept Group also established a framework based on what was effective at other firms. Like all processes, the flow consists of three components—workflow, information/data flow, and decision flow.

The first concern was to map a basic flow. This map illustrates how data and information should be converted into actions and value realization. The Adept Group team then developed a framework for the insight flow. We segmented the information flow into three phases. Refer to Figure CS1.1 below.

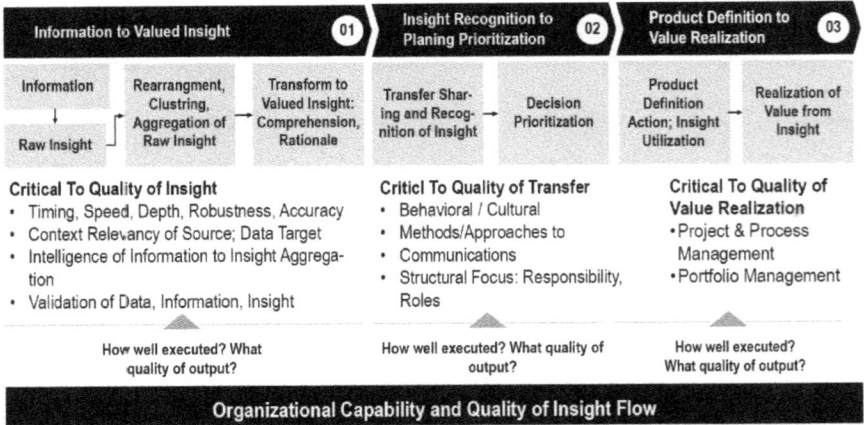

Figure CS1.1 A basic construct of the flow of insights. *It flows from each insights their raw state (sensemaking) to the realization of value.*

The client's challenge was that the information gathered was not translated into insights. The transfer between phases one and two of the flow was not occurring. The client, without realizing it, relied on proximity to facilitate the transfer. This approach was not effective.

The client's size and product complexity necessitated improved information flow throughout the organization. A deliberate process flow addressed this need by linking raw information and insight creation to the value realization efforts.

The market research group and the intellectual property team were collecting a substantial amount of information. These teams, working independently, were synthesizing and transforming the data into insights. Some insights were profound and valuable to the company, while others seemed arbitrary. However, the two groups were not consolidating their insights. Instead, they presented their findings separately during planning meetings with each business unit. There was no established, forward-looking process.

Product development portfolio management was in the early stages of capability maturity for the organization. Stage gate practices, supported by project management for projects, were well-established. Without product line roadmaps, there was no ability to consolidate activities, nor was there a way to connect them to an overall business strategy.

Since the shale oil and hydraulic fracking market was not a targeted segment, no product managers paid attention to it. In business growth vocabulary, this is a "white space opportunity." The opportunity fell into the white spaces on an organizational chart, somewhere between the boxes.

1	2	3
Information to Valued Insight	**Insight Recognition to Planning Prioritization**	**Product Definition to Product Realization**

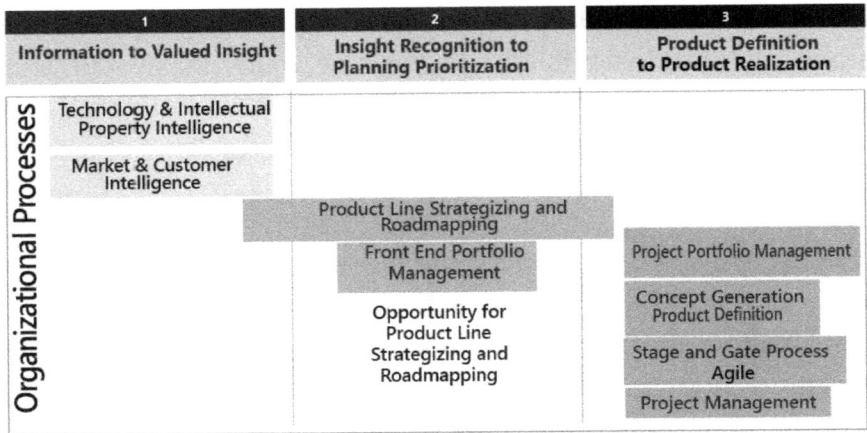

Organizational Processes	Technology & Intellectual Property Intelligence		
	Market & Customer Intelligence		
		Product Line Strategizing and Roadmapping	
		Front End Portfolio Management	Project Portfolio Management
		Opportunity for Product Line Strategizing and Roadmapping	Concept Generation Product Definition
			Stage and Gate Process Agile
			Project Management

Figure CS1.2 Insight and Knowledge Flow

In previous years, the client successfully addressed 'white space opportunities.' Top management mandated the work. The difference was that the client had gone through several years of 'leaning out' their organization. This was a smart response to the economic downturn. There was very little breathing room to pursue such thinking and new endeavors.

The Adept Group team demonstrated to the client leadership how the quality of their work impacted their results. The flow in phases 1 and 3 was good; however, the diagnostic indicated that the flow in phase 2 was poor. Insights were not being recognized as important or prioritized for product planning.

The client's offerings were complex, and the key players were dispersed. The client needed to establish a process that would enable insight to flow from origination through value realization. A critical element missing from the organization's practices was the lack of strategizing and roadmapping for product lines. However, the process needed to encompass a consolidation of strategies and roadmaps. Along with an executive review of all plans, this would reveal all white space opportunities. Hydraulic fracking would have been one of these.

The client carried out the advice. They customized elements of product line strategizing and roadmapping to match their practices. Plus, the client added two event-type methods observed at other companies. After three years of targeting hydraulic fracking, the client's market position went from number six to number four. Sadly, for the client, the entire fracking market then declined.

The Principle of Information and Insight Flow

Well-functioning organizations must proactively guide information flows, especially those with lean practices. This is particularly true when the primary goal of business strategy is to create greater value from new offerings. The flow must be unique to the company and build upon existing practices. Organizations in similar situations should customize a process to create and enhance product line strategic moves and product line roadmaps.

Appendix CASE STUDY 2

Johnson & Johnson

New Platform-levers: How an Archetype Change Can Kill Competition

Before diving into this case study, readers may want to review or reread chapters 9 through 12, which cover product line system archetypes and their impact on performance.

P latforms can be powerful in product line strategies. They can lead to business success or failure. This case study shares insights about new platform-levers and how they can transform a product line system's archetype. It illustrates how new platform-levers, different from industry norms, can disrupt an industry. However, executing new platform-levers effectively requires strategic and creative thinking, along with a significant amount of courage.

This case demonstrates an industry disruption that occurred nearly a decade before Clayton Christensen introduced the concept of disruptive innovation.

Background

Soft contact lenses presented a very interesting business opportunity. The key players in the early to mid-1980s were Bausch and Lomb and CibaVision. By purchasing a small firm, Johnson & Johnson also became a player, albeit a very small one. The industry's technological advances looked like a PhD curriculum in polymer science.

Industry players used polymers such as polyhydroxyethyl methacrylate, hydrogels, siloxanes, and polyethylene glycol. Once formed into a lens, attributes such as oxygen permeability, rigidity, fit, and comfort were important.

Over the years, competitors raced to outperform one another on their lens' molecular makeup. As a result, proprietary polymer formulas became each company's product line platform-lever. All emphasized quality and performance.

At that time, soft lenses had completely replaced hard lenses. Price points ranged from $300 to $500 per pair. Additionally, contact wearers needed to purchase bottles of solutions to clean and store their lenses overnight. Wearing these lenses for more than 18 to 20 hours was not permitted according to FDA-regulated prescriptions.

Competitors could utilize molds or high-precision lathes to manufacture soft lenses. The specific shape and thickness of each lens placed high demands on the manufacturing process. However, the weight of the polymer was only a fraction of an ounce, with water being the distinguishing factor. Consequently, the variable cost for each lens was nearly zero when compared to the selling price.

Johnson & Johnson was establishing a footing in the industry during the mid-1980s. At that time, a small firm was marketing a technology known as 'Danlens.' This technology was cited as:

> "unique multi-patented process which enabled a company to mold lenses in a continuous soft state. This process created a new standard of precision, removing the hydration distortions common to traditional hard-state manufacturing. The result was a high-quality, virtually 100% repeatable and reproducible lens, combining excellent vision with comfort and convenience."

Only J&J showed interest. J and J had a tiny market share. They needed something new to compete with the dominant players.

In 1987, J&J acquired the Danlens' technology and revealed their plans for it. The technology included one other feature that was not directly mentioned in the write-up: it was highly scalable. J&J recognized this potential and quickly capitalized on it. They committed almost $1 billion (equivalent to nearly $3 billion in 2023) to scale the technology. The scale reached levels never before seen in the industry. J & J's production capacity was several magnitudes larger than the combined capacity of other market players.

J&J understood how they could reduce manufacturing costs to unprecedented levels. So low that lens wearers could discard their J and J contacts at the end of each day. Users need not worry about cleaning and storing their lenses. Furthermore, the capacity J and J envisioned would supply over 70% of the market. They calculated that this market share would leave insufficient volumes for competitors to scale up and compete. The approach would succeed if lens wearers disposed of their lenses at the end of each day.

J&J's strategic move was brilliant. They flipped the platform-lever from the industry norm of chemical materials to a large-scale molding production platform-lever. This change in system archetype disrupted the industry. The impact was so significant that it shut down the lens-cleaning solution business, a key revenue source for every player. Ultimately, the contact lens disruption forced Bausch and Lomb into bankruptcy and CibaVision to exit the business.

Lessons to Learn.

Most managers who read this case become enthusiastic about J & J's genius. It was a system archetype change that disrupted the industry. However, many managers overlook how Danlens offered the technology to every player in the industry. Notably, every player except J and J rejected it. This outcome is far more common than strategists care to admit.

Often, the same technologies, markets, and opportunities are accessible to all players in an industry. However, the ability to envision new technology within a product line system makes all the difference. When enterprises align every aspect of their business toward a platform-defined system archetype, it requires tremendous courage to change direction. Yet, J and J accomplished this with the Danlens technology. More notably, they did it boldly, in a multi-billion-dollar manner.

There was a strong economic incentive for J&J to take a different approach. They were far from being the market leader when they entered the market, a position that J&J had never tolerated for long. Bausch and Lomb was the market leader, maintaining a robust cash flow business with the solutions they offered. Why would Bausch and Lomb adopt new molding technology when their margins were already incredibly high? Were they bold enough to cannibalize their product line and scale up a production platform-lever that could change their product line system archetype?

The Important Message

Achieving success in a market can lead a company to become complacent and lack strategic foresight.

We also know this problem as the "success trap," a form of management hubris rooted in past success. Overcoming the success trap requires a leadership team to be strong and willing to resist short-term pressures from shareholders.

Not all new platform-levers will be winners. However, it is essential to understand that both opportunities and constraints will influence decisions about your platform-levers, just as they do for your competitors. Product line teams must leverage this understanding to develop effective strategies and execution roadmaps.

The Success Trap

The success trap occurs when a firm places too much emphasis on exploitation investments, even when it needs explorative investments to better adapt to market changes. Exploitation relies on processes that incrementally improve existing knowledge, whereas exploration focuses on pursuing and acquiring new knowledge.

Firms and other organizations that perform well over an extended period often face strong path dependence in exploitative activities, at the expense of exploratory activities with which they have little experience.

For example, in the 1990s, Polaroid's management failed to respond to the shift from analog to digital photography, despite the emergence of digital technology having been apparent since the 1980s. Other notable examples of companies caught in the success trap include Kodak, Rubbermaid, and Caterpillar.

Two PLSM Principles

The first principle that extends from this case is about success. I state it as: *Watch out for success. It's likely to sneak up on you and bite you in the #!#X. Paranoia often turns out to be far more valuable than the euphoric feeling of success.*

The second lesson focuses on technology. More often than not, the technology needed to transform your line already exists. The bigger issue is having the strategic and creative insight that leads to courageous decisions and investments that can make things work. Every company had access to the technology, but only J&J made the move.

Appendix CASE STUDY 3

Dow Chemical

Changing Platform-lever Types

In the movie The Graduate, someone leaned to Dustin Hoffman and whispered, "Plastics." It was tongue-in-cheek, like saying: "Go ahead and lose your soul." Become a proverbial "Organization Man" in the wonderful world of plastics.

Many intelligent individuals embraced the opportunities offered by the plastics industry. However, not all were willing to accept the subservient and uncreative role of an Organization Man. Kurt Swogger from Dow Chemical was one of those exceptional individuals.

This case concerns thinking outside the box about a product line system.

Intermediate Contribution to an End Product

Most people working on new product development emphasize creativity and concept development. They then hope to turn at least one concept into a product. Ideally, this product will contribute to forming a new product line. It is less common for managers to direct creative thought toward reshaping a product line strategy. Kurt Swogger and his team at Dow Chemical accomplished this with their polyolefin line in the 1990s.

Polyolefins represent the largest category of plastics. A few global producers generate enormous volumes each year to meet the needs of various tangible product companies. Each polymer producer invests in reactors that align with the demand they fulfill. Certain technological advancements, particularly in catalysis and additives, have driven improvements in production efficiency as well as the performance of the polymers themselves. When Swogger took on his role at Dow Chemical, customer performance needs were evolving rapidly, and competition was fierce.

Polyolefins is not a monolithic business; several types of polyolefins exist. The two most common are polyethylene and polypropylene. Various molecule-related qualities contribute to each polymer's performance attributes, which can differ even within the same polyolefin types.

For producers, this is both a blessing and a curse. It is a blessing because it allows the creation of unique polymers that outperform competitive alternatives. However, it is a curse because specialty performance products run contrary to the commodity-oriented and capital-intensive nature of the business.

Diffusion Challenges

Plastics, as sold by producers like Dow and Exxon, are not an end product; they are a material used in the final product. End users have two options if they wish to utilize a new polymer: they can either substitute the new polymer for one already in use, or they may need to design a unique final product to leverage the polymer's new attributes. Both choices contribute to the challenges of diffusing new polymers. Kurt Swogger acknowledged that Dow and Exxon were facing this issue.

The substitute path into the market restricts the potential value that new polymer might offer. First, the polymer must be substituted directly for the original polymer. The success of this method is unlikely since the new polymers generally cost more.

The alternative path is to offer unique properties that enable the design of new end-use products. The buyer must create a new product that exploits the polymer's uniqueness. The issue with this second path is that production volume relies on the customer's product entering the market. Significant volumes may take years to achieve.

Specialty versus Commodity

Competition among polymer producers is intense. Each company must maintain its reactors at nearly full capacity to remain profitable. If production exceeds demand, companies auction off the excess volume, which is detrimental to profits.

The challenge is to cycle through polymer variants during production. Each reactor would transition from one polymer grade to the next. The objective is to minimize waste while aligning supply with demand. Planning must ensure that supply does not get too far ahead of demand when companies launch new polymers.

New polymers with enhanced performance have margins that exceed those of commodity polymers. However, producers face the challenge of achieving sufficient production volume to break even on their reactors. For most polymer product lines, the platform-lever functions as a type of production asset. Typically, the platform-lever consists of a reactor paired with a catalyst.

New Platform-Lever, New Product Line System Archetype

Introducing new polymers can require a significant investment in sales and may also result in notable time delays when filling a reactor. Both Dow Chemical and Exxon Chemical faced these challenges, as each sought to introduce a new type of polyolefin based on their innovative metallocene catalyst.

The metallocene catalyst enabled the tailoring of a "bi-modal" distribution of polymer chain lengths. Each 'mode' provided different attributes. Customizing the polymer for specific features allowed producers to align its performance with each customer's needs. Winning new customers was crucial to filling the reactor's production runs. The race to determine how to accomplish this was intense.

Battles regarding intellectual property were heated. Each company invested heavily to find applications for their bi-modal polymers. Dow had several solution-based reactors dedicated to the metallocene catalyst.

Stuck with the Old

Exxon, in contrast, utilized several large, commodity-scale gas-phase reactors. Solution-based reactors were significantly smaller and had a lower capacity than gas-phase reactors. This smaller reactor size allowed Dow to achieve breakeven sooner than Exxon could.

This is where Kurt Swogger comes in. He was the consummate polymer scientist. Swogger recognized the workload this method imposed on development teams. They had to customize a unique polymer for each client. He understood there were many specific demands.

The challenge was that producing a unique polymer could take up to six months. Swogger reasoned that it was no different for Exxon. If Dow wanted to win the metallocene war, it needed to reduce its sales cycle.

A New Flow of Information

Determining how to produce a specific polymer was a complicated task. Dow and Exxon needed to identify the polymer structure that met customer requirements. They accomplished this in their laboratories through intricate trial-and-error methods.

Next, they moved the prototype to a small pilot plant to refine the production scale-up. The polymer could only enter full-scale production after successful trials at the pilot plant. Throughout this process, customers would test the polymer and provide feedback on its suitability and performance.

The results from the lab were not easily adapted to a pilot plant. Furthermore, what emerged from the pilot plant was also not easily scaled to full-scale production. Overall, the entire process resembled more of an art than a science.

The Strategic Move

Swogger believed Dow could win big if it outperformed Exxon in delivering new polymer to customers. However, the industry prioritized gaining scale by utilizing massive product asset platform-levers. To succeed in this new market, Swogger realized, a design-to-order service platform would be necessary.

Companies can conduct much upfront service in design-to-order work before reaping any rewards. Typically, customers do not pay for the service directly. Companies earned revenue from the sale of the product.

The alternative platform-lever was a software model that rapidly determined polymer characteristics and production settings, outpacing competitors. Mr. Swogger needed to decide if the service platform-lever would provide the advantage necessary to gain market share.

Mr. Swogger examined this challenge and concluded that the software model service platform-lever would be powerful. He believed that the new platform-lever would reduce the time required to develop and produce a customer-specific polymer by half. Furthermore, he reasoned that Exxon would struggle to keep up. Changing Exxon's production, sales, and R&D approaches would be extremely challenging. The company's gas phase reactors and operating practices were not compatible with this new approach.

Swogger's analysis suggested the opportunity was more than sufficient to warrant the investment.

The investment in polymer design and production scale-up modeling surpassed typical laboratory modeling. Significant effort was required to connect the laboratory to the pilot plant and full-scale production. The company also adjusted to support this approach. Key managers concentrated on ensuring they fulfilled their speed promise.

Speed-Based Product Development

Within two years, Swogger's novel platform-lever reduced the time between receiving requirements and full production runs by 80%. The new service capability provided Dow with a significant advantage in the polyolefin market.

The product line strategy at Dow became known as "Speed-Based Development." Dow established itself as an early leader in the metallocene-derived polymer market. In 2003, the Product Development and Management Association awarded Dow its prestigious Innovator of the Year award.

But Situations Change

But by 2005, Dow abandoned its novel platform-lever. Instead, the company switched to a traditional production platform-lever and continued utilizing the model to support production.

As the bi-modal polyolefins market matured, the focus shifted from speed and market diffusion. Large gas-phase reactors dominated the market due to their advantages in scale and cost-effectiveness. Exxon gained market share. In 2000, Dow recognized the signs and acquired Union Carbide Corporation. Among the assets acquired were several large gas-phase polyolefin reactors.

Takeaways for Product Line Systems Management

Driving faster diffusion disrupted the polyolefin market. Dow created a new platform-lever that transformed the product line's archetype to incorporate a valuable service. However, the disruption, while beneficial to Dow, was short-lived. Once Dow had established a new molecular distribution for a customer's end product, Exxon could step in with its commodity-oriented product line and undercut Dow's pricing.

Although numerous metallocene-related intellectual property disputes took place between Dow and Exxon, the most significant clash emerged when one product line archetype faced off against another archetype. Initially, Dow's new archetype created a disruption, but over time, Exxon's traditional commodity-oriented archetype prevailed.

Appendix CASE STUDY 4

Apple

Chain-link Advantage

Steve Jobs would be proud. His company is arguably one of the most successful in history.

Numerous articles have been published praising the strengths of each of Apple's chain-links. One article might commend the advantages of its retail strategy, highlighting how the company outperforms all retailers in sales per square foot, while another recognizes Apple's expertise in its supply chain. Still, other articles laud the company's exceptional "i" branding or acknowledge its intellectual property strategy. Most pundits would agree that Apple's product line strategies are outstanding and significantly contribute to its success.

Each of these chain-links is powerful. (see Chapter 14) However, while we appreciate Apple's products, its success stems from the unique combinations of all its chain-links. It is not just due to its product line strategy.

The synergy Apple gains across its chain-links makes it nearly impossible for competitors to replicate. In its overall business strategy, Apple's chain-link synergy creates an economic moat that insulates it from rivals. Competitors may match or exceed one or two of Apple's strategic links, but they find it impossible to copy all of Apple's links consistently. For example, a major competitor like Samsung may match or even surpass Apple's iPhone product line strategy. However, Apple remains the market leader because Samsung continually falls short on other chain-links.

The biggest threat to Apple's chain-link prowess comes from the Chinese market. In China, Apple's functional chain-links don't create synergy to the same extent as they do in the West. Its retail, supply chain, content delivery, and intellectual property links are not as effective there as they are in the West. As a result, stock rating firm Morningstar lowered Apple's economic moat rating (as discussed in Chapter 3). This downgrade is certainly a major concern for Apple, given the size of the Chinese market and its influence across Asia.

Appendix CASE STUDY 5

Keurig® Coffee

Emergence and Thinking You Know Best

L et's look at a simple example of what can happen even to a young company that didn't grasp the complex multi-product nature of its offerings. That company is Keurig, the coffee brewer manufacturer. Keurig has a very strong and positive culture that emphasizes entrepreneurial success. Most people who work for Keurig are, and continue to be, very proud, and they cringe at discussing mistakes. However, when Keurig attempted to innovate using their "next big thing" entrepreneurial approach, it worked against them.

How Keurig Missed the Boat

Keurig is the company that invented the single-serve coffee brewer, which it launched in 1998. The machine appeared in its target market—corporate offices—shortly after the launch and transitioned into the consumer market a few years later. For many years, Keurig dominated this fast-growing market, virtually monopolizing this new way to produce individual cups of coffee with less mess.

But the company recognized that the patent on its single-serve K-Cup pods (the key platform-lever in its offering) was set to expire in 2012. To maintain its growth trajectory, Keurig needed to broaden its product family with a new platform-lever. The responsibility was therefore placed on Keurig's designers to create something new—the company's next big thing. Sure enough, they delivered. It was the stylish Keurig 2.0—a hot beverage brewing system that featured the added capability to brew a large canister of coffee.

Keurig's CEO Brian Kelley believed they had hit a home run. "Our new Keurig 2.0 system is the next innovative step for our hot beverage brewing system," he stated in 2014. Unfortunately, consumers disagreed. Sales were low. Complaints poured in. The market balked. What was happening? "Full pots ... [were] ... something we bought Keurig machines not to have to do in the first place," wrote Daniel Kline for Motley Fool in April 2015.

The next big thing turned out to be not so big!

Where Did Keurig Go Wrong?

Could this stumble on the road have been avoided? Yes, it could have. In fact, a strong product line strategy ensures a match between product attributes and what customers truly desire.

The discipline of defining the right product line strategy with insights about a new platform-lever would have identified the mistake of prematurely forcing the "next big thing." It would have focused on the critical nature of market segmentation, both as a noun and a verb. It would have aligned with the nuances of the customer's true wants, needs, and, most importantly, desired "outcomes." This is what I refer to as "attribute positioning," and it is fundamental to effective product line strategies. See Appendix 7.

Product line attribute positioning involves the intentional targeting, development, and delivery of product or service features. This approach aims to establish a market position within a segment. It aligns product attributes with customer needs, desires, and intended use outcomes.

Attribute positioning is essential for effective product line strategies. In this situation, it would have encouraged both developers and marketers to collaborate and explore alternative options. Without strategic attribute positioning, Keurig squandered a significant amount of time and money. While the 2.0 version was available for a period, it was not a new offering; rather, it was simply a revised version of the 1.0 tailored to the original platform-lever.

Many people who hear the Keurig story immediately emphasize the need for improved customer research. Yes, this is true. However, another aspect of the complex nature of product lines relates to Keurig's strong culture and the behavior it fostered. A very positive, can-do attitude was fueled by the company's immense success up to that point. And who can argue with success?

The challenge is to change behaviors and judgments to enable well-considered product line attribute positioning. It should not be merely about reacting to the next big thing. Keurig reminds us that success-induced hubris can lead companies into significant problems.

ENDNOTES, GLOSSARY, INDEX, AND THE AUTHOR

ENDNOTES

Chapt	Ref #	Citation
Forward	1	Kauffman, Stuart. "The 'Adjacent Possible'—and How It Explains Human Innovation." TEDxRainier (video). November 2014.
Forward	2	Woods, Audrey. "The Death of Moore's Law: What it means and what's filling the gap going forward." MIT Computer Science and Artificial Intelligence Laboratory.
Forward	3	The Shift from Models to Compound AI Systems | Berkley Artificial Intelligence Research | 18 Feb 2024 | Multiple Authors | https://bair.berkeley.edu/blog/2024/02/18/compound-ai-systems/
Forward	4	OpenAI reports that they reduced ChatGPT's cost of delivering a standard output unit by ninety-nine percent in just two years.
Forward	5	Musk Says AGI 2026; Matthew Berman Dec 2024 https://m.youtube.com/watch?v=49bxxSns1jk
1	6	How Domain-Specific AI Agents will Shape the Industrial World in the next 10 years, by Christopher Nguyen, CEO AITOMATIC at the 3rd Annual Industrial AI Conference, Stanford University.
1	7	O'Reilly III, Charles A., and Michael L. Tushman. "The Ambidextrous Organization." Harvard Business Review 82, no. 4 (April 2004): 74-81.
1	8	Christensen, Clayton M. The Innovator's Dilemma: When New Technologies Cause Great Firms to Fail. Boston: Harvard Business School Press, 1997.
1	9	Eric Ries, The Lean Startup: How Today's Entrepreneurs Use Continuous Innovation to Create Radically Successful Businesses. Currency, 2011.
1	10	Pinchot, Gifford (1985). Intrapreneuring: Why You Don't Have to Leave the Corporation to Become an Entrepreneur. New York: Harper & Row.
1	11	O'Connor, G., Rice, M., and colleagues. "Radical Innovation: How Mature Companies Can Outsmart Upstarts," Harvard Business School Press, 2000.
1	12	Ralph-Christian Ohr | LinkedIn: https://www.linkedin.com/in/ralphchristianohr/ Web: https://dual-innovation.net/
1	13	Enduring Ideas: The three horizons of growth, Steve Coley, McKinsey Quarterly Dec 2009.
1	14	Blank, Steven. "Why Companies Do 'Innovation Theater' Instead of Actual Innovation", Harvard Business Review, February 1, 2019

Chapt	Ref #	Citation
1	15	Christensen, Clayton M., and Michael E. Raynor. The Innovator's Solution: Creating and Sustaining Successful Growth. Harvard Business Review Press, 2003.
1	16	O'Connor, Gina Colarelli. Grabbing Lightning: Building a Capability for Breakthrough Innovation. San Francisco: Jossey-Bass, 2008.
1	17	Mattes, Frank. Now and New: How Companies Can Use Their Capabilities for Scaling. Waldkirch, Germany: Now and New, 2024.
1	18	Blank, S. (2017, October 30). Why GE's Jeff Immelt Lost His Job: Disruption and Activist Investors. Harvard Business Review.
1	19	Steve Blank: McKinsey's Three Horizons Model Defined Innovation for Years. Here's Why It No Longer Applies.
1	20	DeepMind's director, Dr. John Jumper, and Demis Hassabis were awarded the Nobel Prize in chemistry.
1	21	Volkswagen's crisis: How can Europe's car industry survive? : Deutsche Welle: https://amp.dw.com/en/volkswagens-crisis-how-can-europes-car-industry-survive/a-70231806
2	22	David Graeber, Bullshit Jobs, A Theory 2018, Simon & Shuester.
2	23	Walo, S. (2023). 'Bullshit' After All? Why People Consider Their Jobs Socially Useless. Work, Employment and Society, 37(5), 1123-1146. https://doi.org/10.1177/09500170231175771
2	24	Doctorow, Cory (January 23, 2023). "The 'Enshittification' of TikTok". Wired. Condé Nast.
2	25	Single Vs. Multi-Product Thinking, The Adept Group, https://adept-plm.com/single-versus-multi-product-thinking/
2	26	Product Lines as System, The Adept Group, https://adept-plm.com/product-lines-as-systems/
2	27	Roger Martin Strategy; Roger Martin is co-author with P&G CEO Alan Lafley of 'Playing to Win: How Strategy Really Works', 2013.
2	28	S-Curve: Foster, Richard N., Lawrence Linden, Roger Whitely, and Alan Kantrow. "Improving the Return on R&D - I," Research Management, Vol. 28, No. 1 (January-February 1985), pp. 12-17.
2	29	Wright Laws | Hirschmann, Winfred B. (1964-01-01). "Profit from the Learning Curve". Harvard Business Review. No. January 1964.
2	30	Michael Porter -Differentiation Porter, Michael E., "Competitive Advantage". 1985, Ch. 1, pp 11-15. The Free Press. New York.
2	31	Moskowitz, Howard R. "Product Differentiation: A Key to Successful Product Development." Food Product Design, vol.. 4, no. 11, 1994, pp. 60-69.
2	32	Snowden, Dave J. "As through a glass darkly: A complex systems approach to futures." Handbook of Futures Studies, Edward Elgar

Chapt	Ref #	Citation
		Publishing, 2024, pp. 48-65. | https://thecynefin.co/knowledge-mapping-3-of-3/
3	33	Economic Moats Matter: Here's the Evidence, Holt and Lane, Sep 2, 2017 https://www.morningstar.com/portfolios/economic-moats-matter-heres-evidence
3	34	April 2023 internal Google memo, titled "We Have No Moat, And Neither Does OpenAI," argued that both Google and OpenAI lacked a sustainable competitive advantage in artificial intelligence, particularly due to the rapid advancements in open-source models.
3	35	Customer-centric Innovation Design Thinking: Integrating Innovation, Customer Experience, and Brand Value; Thomas Lockwood and Edgar Papke, 2010.
4	36	What is Product Line Velocity?, The Adept Group, https://adept-plm.com/what-is-product-line-velocity/
6	37	What is a Product Line and Why Is It So Important?, The Adept Group, https://adept-plm.com/product-line-definition-1/
6	38	"How Ice Cream Kills! Correlation vs. Causation https://m.youtube.com/watch?v=VMUQSMFGBDo by Decision Skills"
6	39	Wikipedia: Complex adaptive system; https://en.wikipedia.org/wiki/Complex_adaptive_system
6	40	Snowden, Dave J. "As through a glass darkly: A complex systems approach to futures." Handbook of Futures Studies, Edward Elgar Publishing, 2024, pp. 48-65. | https://thecynefin.co/knowledge-mapping-3-of-3/
7	41	6Sigma.us. (2024, October 10). Lean Flow: The Ultimate Guide to Streamlining Your Business Processes. 6Sigma.us.
8	42	Reinertsen, Donald G. The Principles of Product Development Flow: Second Generation Lean Product Development. Redondo Beach, CA: Celeritas Publishing, 2009.
8	43	Cooper, R. G. (1985). The NewProd System: The Industry Experience. Journal of Product Innovation Management, 2(1), 13-27.
8	44	O'Connor, Paul. "A Toolbook of Product Development Portfolio Management Capability Model." International Journal of Technology Management 18, no. 4 (2003): 25-37.
9	45	O'Connor, Paul. The Profound Impact of Product Line Strategy: Your Guide to Highly Productive Product Line Management. APT Publishing, 2018.
10	46	Krueger, C. W. (2010). Systems and Software Product Line Engineering with the SPL Lifecycle Framework. In Proceedings of the 14th International Software Product Line Conference (SPLC) (pp. 34-41). Springer, Berlin, Heidelberg.

Chapt	Ref #	Citation
10	47	Biryulin, S. (2022, September 18). Forecast and Foresight - What's the Difference? Medium.
11	48	Rutherford, A. (2019). The Elements of Thinking in Systems: Use Systems Archetypes to Understand, Manage, and Fix Complex Problems and Make Smarter Decisions. Independently published.
11	49	What Is a Business Model? by Carol M. Kopp, Updated June 13, 2025, Investopedia
11	50	What is Conway's Law?, Microsoft 365 Life Hacks, June 28, 2024 | https://www.microsoft.com/en-us/microsoft-365-life-hacks/organization/what-is-conways-law
11	51	The Mirroring Hypothesis: Theory, Evidence and Exceptions, by Carliss Y. Baldwin, Harvard Business School Working Paper Series 2016 Page 75
12	52	Bower, J. L., & Christensen, C. M. (1995). Disruptive Technologies: Catching the Wave. Harvard Business Review, 73(1), 43-53.
13	53	Juarrero, A. (2023). Context Changes Everything: How Constraints Create Coherence. The MIT Press.
13	54	Snowden, D. (2023, March 14). Dave Snowden - Estuarine mapping - Talk - Agile Tuesday [Video]. YouTube. https://www.youtube.com/watch?v=-SITXkTFHVw
13	55	Greg Lukianoff (2018). The Coddling of the American Mind: How Good Intentions and Bad Ideas Are Setting Up a Generation for Failure. New York City: Penguin Press.
14	56	Rumelt, R. P. (2011). Good Strategy Bad Strategy: The Difference and Why It Matters. Crown Business.
14	57	Farley, J. (2025, January 15). Ford CEO's brutal honesty: We can't compete with Tesla [Interview]. Fully Charged Show. YouTube. https://www.youtube.com/watch?v=JdId8tXiJIo
14	58	Sandy Munro in his Munro Live YouTube vlog. He compares car design and manufacturing methods across automakers
14	59	Paul Hobcraft on Ecosystems and Innovation, https://ecosystems4innovating.com/
14	60	The International Organization for Standardization (ISO) is a global body that develops and publishes international standards to ensure quality, safety, efficiency, and interoperability across various industries.
15	61	Outcome-Driven Innovation: JTBD Theory in Practice, Tony Ulwick, July 2017 https://jobs-to-be-done.com/outcome-driven-innovation-odi-is-jobs-to-be-done-theory-in-practice-2944c6ebc40e
15	62	Christensen, Clayton M., Karen Dillon, Taddy Hall, and David S. Duncan. Competing Against Luck: The Story of Innovation and Customer Choice. Harper Business, 2016.

Chapt	Ref #	Citation
15	63	Crawford, C. Merle. "Defining the Charter for Product Innovation." Sloan Management Review, vol. 22, no. 1, 1980, p. 3.
15	64	Roger Martin Playing to Win: How Strategy Really Works; Roger Martin is co-author with P&G CEO Alan Lafley of 'Playing to Win: How Strategy Really Works', 2013
15	65	O'Connor, Paul. The Profound Impact of Product Line Strategy: Your Guide to Highly Productive Product Line Management. APT Publishing, 2018.
16	66	Martin, Roger. "A Strategic Framework for Growth." Medium, 22 July 2024,
17	67	Githens, Greg. "Insights Are the Secret Sauce of Strategy." Leading Strategic Initiatives, 27 Nov. 2019, https://leadingstrategicinitiatives.com/2019/11/27/insights-are-the-secret-sauce-of-strategy/.
17	68	University of Michigan Professor Karl Weick's 1995 seminal publication, Sensemaking in Organizations.
17	69	Ryder, Mike; Downs, Carolyn (November 2022). "Rethinking reflective practice: John Boyd's OODA loop as an alternative to Kolb". The International Journal of Management Education.
17	70	Thurlow, Nigel; An Explanation of OODA and Its Relevance To Complex Systems. 2021; https://www.youtube.com/watch?v=awGcZ5KIGAM
17	71	Malcolm Gladwell: The Tipping Point; Blink; Outliers; What the Dog Saw.
17	72	Safe to fail probes, Cynefin Company; https://cynefin.io/wiki/Safe_to_fail_probes/
17	73	Githens, Greg. How to Think Strategically: Sharpen Your Mind. Develop Your Competency. Contribute to Success. Maven House Press, 2019.
18	74	Crawford, C. Merle. "Defining the Charter for Product Innovation." Sloan Management Review, vol. 22, no. 1, 1980, p. 3.
19	75	Kate W. Issac; "Why Distributed Leadership is the Future of Management"; April 19, 2022; MIT Management
20	76	Northouse, Peter (2007). "Contingency Theory". Leadership: theory and practice. Thousand Oaks: SAGE Publishing.
20	77	Rao, Jay. W.L. Gore: Culture of Innovation. Case No. BAB698, Babson College, 2012.
21	78	"Tesla will launch unsupervised driving in June, Musk says"; 29 January 2025, The Verge.
21	79	Estuarine framework, Dave Snowden, Cynefin Company: https://cynefin.io/wiki/Estuarine_framework

Chapt	Ref #	Citation
21	80	Isaacson, Walter. Elon Musk. Simon & Schuster, 2023.
23	81	Mike McGrath's (the M in PRTM consulting, now part of PwC) PACE process.
23	82	Cooper, R. G. (2017). Winning at New Products: Creating Value Through Innovation (5th ed).. Basic Books.
23	83	Deschamps, J.-P., & Nelson, B. (2014). Innovation Governance: How Top Management Organizes and Mobilizes for Innovation. Jossey-Bass.
24	84	Anderson, P. W. (1972). More is Different. Science, 177(4047), 393-396. | Also: "More is different": Why reductionism fails at higher levels of complexity https://bigthink.com/13-8/reductionism-fails-complexity/
26	85	O'Connor, Paul. "A Toolbook of Product Development Portfolio Management Capability Model." International Journal of Technology Management 18, no. 4 (2003): 25-37.
26	86	Lewin, K. (1947). Frontiers in group dynamics: Concept, method and reality in social science; social equilibria and social change. Human Relations, 1(1), 5-41.
27	87	Building innovation ecosystems: Accelerating tech hub growth; McKinsey and Company, Feb 2023 https://www.mckinsey.com/industries/public-sector/our-insights/building-innovation-ecosystems-accelerating-tech-hub-growth
27	88	Githens, Greg. "Insights Are the Secret Sauce of Strategy." Leading Strategic Initiatives, 27 Nov. 2019, https://leadingstrategicinitiatives.com/2019/11/27/insights-are-the-secret-sauce-of-strategy/.
28	89	The Invention of the Post-it® Note, June 2020 https://www.invent.org/blog/trends-stem/who-invented-post-it-notes
29	90	Fusion Strategy, Harvard Business School Publishing, 2024, Govindarajan and Venkatraman.
29	91	Sherry, Yash, and Neil C. Thompson. "How Fast Do Algorithms Improve? [Point of View]." Proceedings of the IEEE, vol. 109, no. 11, 2021, pp. 1768-77. https://doi.org/10.1109/JPROC.2021.3107219
29	92	Digital Transformations and Product Line Performance, June 2021, https://adept-plm.com/digital-transformations/
29	93	Singh, V., & Willcox, K. E. (2018). Engineering Design with Digital Thread. In 2018 AIAA/ASCE/AHS/ASC Structures, Structural Dynamics, and Materials Conference (pp. 1-15). American Institute of Aeronautics and Astronautics.

Chapt	Ref #	Citation
29	94	Cooper, Robert G. 2024. The Artificial Intelligence Revolution in New-Product Development." IEEE Engineering Management Review 52(1), Feb. 195-211.
29	95	Andy Jassy on using generative AI in software development at Amazon; August 2024; https://werd.io/2024/andy-jassy-on-using-generative-ai-in-software-development-at
29	96	Dunning, D., & Kruger, J. (1999). Unskilled and unaware of it: How difficulties in recognizing one's incompetence lead to inflated self-assessments. Journal of Personality and Social Psychology, 77(6), 1121-1134.
29	97	How TSMC convinced Apple it would be a trustworthy partner; Fortune Magazine, February 10, 2023.
29	98	von Hippel, E. (1986). Lead Users: A Source of Novel Product Concepts. Management Science, 32(7), 791-805.
30	99	Product Development and Management Association; https://www.pdma.org/
30	100	Outcome-Driven Innovation: JTBD Theory in Practice, Tony Ulwick, July 2017 https://jobs-to-be-done.com/outcome-driven-innovation-odi-is-jobs-to-be-done-theory-in-practice-2944c6ebc40e
30	101	Alex Osborn; Applied Imagination: Principles and Procedures of Creative Problem Solving. New York: Charles Scribner's Sons, 1953.
30	102	"The Gang That Couldn't Think Straight"; INC Magazine, October 1986. (This article is about an innovation service firm I worked at in the mid-1980s. Gifford Pinchot, the creator of Intraprenuership, also worked there prior to writing his book).
30	103	Moore, G. A. (1991). Crossing the Chasm: Marketing and Selling High-Tech Products to Mainstream Customers. Harper Business.
A.1	104	Learning Curves and Wright's Law, by Tom Connor, Medium, Feb 26, 2022; https://medium.com/10x-curiosity/learning-curves-and-wrights-law-744b85b897a2

GLOSSARY OF TERMS

Term	Definition
Accelerating Change Race	The intense, accelerating change that companies must confront, demanding continuous enhancement of offerings for efficiency and facilitation of change to address new and future demands. It underscores the need to redefine product innovation and management.
Adjacent Possibles	Innovation can introduce an extraordinary change, but the changes are limited to the combination of things we already know—the adjacent possibles. If we know more things, there are more combinations possible. AI-supported knowledge increases the adjacent possible (and innovation possibilities) exponentially.
Affordance	The environmental and systemic enablers that invite and aid individuals in generating insights and making sense of complex systems.
Agent	A general term for components within a product line system, including key parts, enablers, constructors, subsystems, and actors. Interactions among these agents dictate the system's performance and its ability to adapt.
AI Enablement	The integration of AI, machine learning, and analytics into product line systems to enhance efficiency and adaptability.
Ambidextrous Organization	An organization designed to balance exploration of new innovations with exploitation of existing business strengths.
Archetype Disruption Theory	A theory stating that disruption in an industry occurs only when an innovation introduces a new system archetype, regardless of the size of the originating company.
Architecture Change Levels	Four levels of system adaptability: attribute updates, product customization, platform-lever advancements, and composable system design.
Assemblage Mapping	The creation of a mapped plan to add, modify, improve, or eliminate system agents and parts to improve the product line system's structure and performance.

Term	Definition
Attitudinal Mindset	Refers to people's behaviors, attitudes, and understanding, which some experts argue need to change to address challenges like the Accelerating Change Race. The book argues that changing the context is more important than just changing mindsets.
Automated Reasoning	An AI enhancement that, when integrated with Retrieval Augmented Generation, contributes to the accelerating pace of change by enabling AI agents to perform many tasks traditionally done by humans.
Breakthrough Innovation	A term often used to describe radical innovations by large companies, sharing the same core novelty as disruptive innovation but applied to bigger firms.
Bridging Enabler	Temporary systems or tools created to address gaps and enable changes until permanent solutions are developed.
Centers of Excellence	Dedicated groups or departments that gather and expand knowledge on critical topics or technologies for product line advancement.
Chain-Link Strategy	An analogy where each functional strategy within a business unit is viewed as a link in a chain that contributes to the overall strategy.
Change Capacity	The ability of a product line system to adapt and evolve efficiently in response to external and internal changes.
Change Equilibrium Equation	A concept from Kurt Lewin stating that organizations "unfreeze" and embrace change when the perceived and real gains, plus the perceived losses from not changing, are greater than the perceived and real economic and psychological costs of changing.
Complex Adaptive System	A system composed of interconnected agents or parts that adapt and self-organize in response to internal and external influences, making outcomes emergent rather than predictable.
Composability	A feature of a product line's design (often enabled by Product Line Engineering) that allows for the easy composition of

Term	Definition
	new offerings using predefined modular blocks, enhancing flexibility and Change Capacity.
Compound AI Systems	More sophisticated AI systems that integrate facilitated prompts, reasoning, planning, and action-taking, transforming from simpler large language models.
Constructor	A type of system enabler that directly creates or modifies system parts, agents, and flows; often using AI or machine learning.
Context/System Substrate	The slice of the infinitely large system (the universe of everything) on which you choose to focus. The foundational shift from single-product orderly systems to multi-product complex systems. The product line is a substrate within a business unit, within a corporation, within an industry, within a society, and economy.
Corporate Innovation (CI)	A deliberate approach by large companies to drive organic growth through major innovations, distinctly separated from core offerings.
Corporate Venturing	Investments by large companies in startups to acquire new product lines or system components for growth and innovation.
Cynefin Framework	A decision-making framework that helps understand and manage systems based on their level of uncertainty, guiding appropriate strategies for complex systems; Credit to Dave Snowden and his firm, The Cynefin Company.
Digital Thread	The integration of data and ecosystems into a unified model enables comprehensive (as full as possible) system visualization and management.
Distributed Leadership	A leadership model where decision-making and initiative are spread across roles and functions, promoting self-directed efforts.
Disruptor Theory	A theory positing that disruption of an industry occurs when an innovation is based on a product line system archetype that is new to that industry.

Term	Definition
DOJO (Tesla's)	A custom-designed, billion-dollar supercomputer developed by Tesla to process vast amounts of visual data for constructing autonomous driving algorithms and the AI core of humanoid robots. It serves as a "Constructor" for the product line system.
Dunning-Kruger Effect	A cognitive bias where people with limited expertise in a topic tend to overestimate its potential when first learning about it.
Ecosystem	The external-to-the-line collaborations and partnerships, purposely managed to boost the product line's performance, may be external to the company or internal functions and departments within the company.
Efficiency Inertia	An organizational force created when companies double down on core product efficiency, which resists change and prevents core businesses from pursuing significant innovations. It also generates a greater need for impactful Corporate Innovation.
Efficiency-Change Dilemma (New)	The same as the old challenge, but greatly amplified by accelerating change, requires changing the core product offerings. The two orientations, efficiency and change, are not separate or independent of one another.
Efficiency-Change Dilemma (Old)	A challenge faced by companies in balancing the need to optimize efficiency in core products while embracing major innovations to adapt to rapid market and technological changes. These activities are separated organizationally.
Enabler Roadmap	A plan to implement and improve system agents, tools, practices, and processes supporting the product line system's advancement.
Enabler-Less System	A product line system that lacks the necessary tools, practices, or processes to sustain and enhance performance.
Enabling Constraint	A boundary or guideline in a complex system that directs actions and decisions toward desirable outcomes. Strategies and charters are enabling constraints in product line systems.

Term	Definition
Enshittification	A term used by Cory Doctorow to describe how online products and services can decline in quality, stemming from the ease of changing how a platform allocates value and holding buyers and sellers hostage while raking off increasing value.
Estuarine Mapping	A methodology developed by Dave Snowden that helps teams categorize and plot constraints based on the time and effort needed to address or avoid them, leveraging a complex system's natural adaptation to navigate roadblocks by identifying the path of least
Flow Chaos	The random flow of work, information, decisions, and outcomes within a product line system, manifesting as disruptive and inadequate processes.
Flow Coherence	The state achieved by intentionally adjusting and matching product line flows (work, information, decisions, outcomes) to imply un
Flow Enablers	System agents, such as tools, practices, and processes that enhance system flows (workflows, decision flows, etc.), such as agile methodologies or portfolio management software.
Flow Resonance	A state where all flows (workflows, information flows, decision flows, and outcome flows) in a product line system amplify each other, enhancing outcome emergence and strategic advancements.
Futurecasting	A team's best speculation about outcomes for a complex system concerning internal and external influences and scenarios, often relying on qualitative aspects and presenting the most likely scenarios with narratives explaining the rationale. It differentia
Governance	The oversight mechanism in product line systems management, focused on enabling and guiding system performance and adaptability.
Heuristics	Rules of thumb or mental shortcuts derived from experience to simplify decision-making in complex systems.

Term	Definition
Innovation Charter	A conceptual object that captures insights and serves as a guide for creating potential strategic moves within a product line.
Innovation Theater	Superficial innovation efforts, such as labs or hubs, that focus on appearance rather than impactful results.
Insights	Bits of new knowledge and understanding that combine with existing knowledge to reduce uncertainty and inform strategic moves.
Jobs-to-be-Done (JTBD)	A theory of customer needs based on the idea that people "hire" products to complete a "Job" or achieve a desired outcome. It is used to assess customer satisfaction and identify opportunities.
Lattical Organizational Structure	A hybrid organizational structure that combines elements of both hierarchical and lattice structures, aiming to balance efficiency and change by allowing individuals to self-direct while maintaining some formal reporting lines.
Moore's Law	The historical observation that CPU chip performance doubles approximately every eighteen months. The book argues that AI's advancements are making this too slow and even irrelevant.
Multi-Product Thinking	A modern approach that manages a set of related products (product lines) as interconnected systems, leveraging synergies, shared resources, and strategic advancements to improve overall performance and adaptability. Contrasts with Single-Product Thinking
Negative Innovation Loop	An ongoing systemic feedback loop where efforts to increase efficiency in core products reduce the organization's ability to innovate, which in turn pressures core products to support innovation, perpetuating the underperformance of corporate innovation a
Noun-Based Segments	Traditional market segmentation categories such as geography, demographics, or psychographics.
OODA Loop	A framework (Observe, Orient, Decide, Act) developed by military strategist Col. John Boyd, valuable in complex

Term	Definition
	systems for understanding situations through rapid iterative cycles, especially in uncertain environments.
Outcome Flow	The combined results of workflows, decision flows, and information flows in a product line system.
Outcome Velocity Vector	A key indicator of a product line's performance across customer satisfaction, cash flow, and competitiveness. The vector has a direction and a magnitude (length).
Platform-Lever	A critical component of a product line system that provides "leverage" (faster development, greater attribute performance, decreased per-unit costs) across multiple products within a line. It plays a central role in a product line's architecture and strategy.
Platform-Lever Advancements	Strategic innovation to advance or transform a platform-lever; like Apple's M-series chips, designed to provide leverage across multiple products and generations, boosting performance and synergies within a product line.
PLSM (Product Line Systems Management)	A multi-product complex systems management approach emphasizing strategic moves, Outcome Velocity improvements, and Change Capacity.
PLSM Praxis	The combination of theory and practice in Product Line Systems Management.
Product Line	A set of related products managed as a strategic unit to maximize competitive advantage, efficiency, and customer satisfaction.
Product Line Archetype	The foundational model or structure that defines the system behaviors and flows of a product line. Common system archetypes have common behaviors. Product line archetypes are dependent on platform-lever types and combinations.
Product Line Assemblage	The full system of components, agents, and interactions that form a product line, including products, platform-levers, technologies, market segments, and strategic moves.

Term	Definition
Product Line Change Theory	A theory positing that product line systems change more rapidly and efficiently when requisite changes to enablers and ecosystems, at their finest level of granularity, accompany key part changes.
Product Line Congruency Theory	*A theory positing that to achieve superior performance of a product line and the business in which it operates, the product line system archetype must align with the business model type.*
Product Line Enablement Pyramid	A framework illustrating how a product line's structure, strategy, and system components align to enhance performance (efficiency and change).
Product Line Engineering (PLE)	An engineering approach that enhances product development flexibility through modularity and predefined components, creating a feature library for faster customization and innovation.
Product Line Outcome Velocity	A measure of a product line's progress across three dimensions: cash flow, competitiveness, and customer satisfaction, represented as a vector showing the direction and magnitude of change over time.
Product Line Responsiveness	The ability to act quickly and effectively when meaningful insights emerge, driving value creation.
Product Line Roadmapping	An enabler - A tool for planning, aligning, and tracking the components, flows, and initiatives within a product line system.
Product Line Stack	The stack-like assemblage of key product line system parts, connecting technologies to customer needs through platform-levers and products. A product line stack is a static view of a product line roadmap.
Product Line System Audit	A systematic evaluation of the components, agents, and interactions within a product line system to identify improvements.
Product Line Systems	A powerful framework that represents a significant shift in context, guiding teams to navigate and leverage uncertain and dynamic situations by understanding and managing product lines as complex, multi-product systems.

Term	Definition
Management (PLSM)	
Retrieval-Augmented Generation (RAG)	An AI capability that enhances Large Language Models (LLMs) within a company's specific context domain by "chunking and embedding" an organization's internal knowledge base (documents, reports, emails).
Ring-Fencing	A strategy to isolate Corporate Innovation activities from a company's core business to reduce interference and risk.
Roadmap for Enablers	A strategic plan outlining current and future states of enablers to support a product line's Outcome Velocity and adaptability. See Assemblage Mapping.
Safe-to-Fail Experiments	Experiments designed to generate insights in complex systems without risking significant harm to the organization or project, especially relevant when the results are uncertain.
Sensemaking	The broader task of stitching together insights and existing knowledge to create meaning and guide decisions in complex systems.
Single-Product Thinking	A traditional approach focusing on the development, management, and optimization of one product at a time, often missing cross-product synergies and strategic gains. Contrasts with Multi-Product Thinking.
Single-Product Thinking	A pervasive, traditional paradigm focused on the development, management, and optimization of one product at a time, often overlooking cross-product synergies and hindering adaptation to rapid change.
Strategic Debt	Postponed or overlooked strategic moves that accumulate, acting as a hidden negative force that can threaten daily operations and increase the riskiness of a situation.
Strategic Move	Actions that align with 'adjacent possibles' to drive growth and improve product line Outcome Velocity.

Term	Definition
Strategic Move Coherency	Ensuring that strategic actions align with system constraints, build on one another, and contribute meaningfully to product line Outcome Velocity.
Strategic Move Stream	A series of interconnected and coherent strategic moves that collectively enhance a product line's Outcome Velocity and growth.
Substrate (Complex System Substrate)	Refers to the area or basis of the system on which it operates and evolves, establishing the context for reflections, and is shaped by the constraints and forces to be navigated.
Success Hubris	A significant issue affecting product line changes, where past success leads to overconfidence and a mindset of "we know best," hindering coherent strategic moves and adaptability.
System Archetype	The underlying structure that influences the behavior patterns of a product line system. See Product Line Archetype.
System Change Capacity	A system's ability to adapt and evolve through modifications to its agents, parts, and flows.
System Constraints	Limitations, boundaries, and hindrances that affect a product line system, its flows, and performance.
System Enablers	Tools, processes, and practices designed to improve a system's performance, flows, or adaptability.
System Flows	The interconnected decision, information, work, and outcome flows within a product line system.
Time-to-Market (T2M)	A traditional Key Performance Indicator (KPI) for product development speed, focusing on how quickly new products are launched. The book argues its correlation with favorable outcomes has diminished due to universal emphasis.
Transformative Shifts	A higher order of strategic moves that involve a change in the product line's system archetype, purposefully altering its architecture and ecosystem, leading to significantly different outcomes and promoting growth.

Term	Definition
Uncertainty Flow	The progression of managing uncertainty in Corporate Innovation from chaotic to complex and eventually toward order.
Unlearning	The process of addressing and diminishing negative forces stemming from entrenched practices, outdated processes, and deeply rooted heuristics and assumptions before a change takes place, also referred to as "unsticking" system agents.
Verb-Based Segment	Market groupings based on clusters of desired outcomes (Jobs-to-be-Done) rather than demographic or geographic categories.
Weak Signal Sensing	The ability to identify and interpret faint signals of change before they develop into stronger ones, allowing teams to derive insights and transform them into effective actions ahead of competitors.
Wright's Law	A principle stating that the cost of producing a product declines by 15% to 20% as cumulative production volume doubles, suggesting that scale improves "bang for the buck." The book notes AI's advancements are challenging its timelessness.

INDEX OF TOPICS

575

577

product line system archetypes, 143–45, 156, 161–62, 256, 326, 372, 395, 501, 505, 529, 534
Product Line System Change Theory, 30
product line system's archetype, 140, 153–54, 159, 181, 405, 467, 479, 501
product line transformation, 153–54, 213, 296, 467
product management, 24, 32–33, 107, 239, 328, 344, 370, 398, 446, 473, 491
product offerings, 15, 32, 214, 288, 292, 316, 332, 341, 350, 380, 494–95
products-in-development (PID), 109–10, 123, 207, 211, 271, 480
products-in-the-market (PIM), 109–10, 123, 207, 271, 370, 480
project portfolio management, 107, 300, 306, 384, 401, 444
project speed, 45, 49, 104, 234
PRTM, 306, 525

ABOUT THE AUTHOR

Paul O'Connor - Advisor, Teacher, Facilitator, Author

For decades, Paul O'Connor has been driven by a single focus: to improve and simplify innovation within mid-size and large companies. His work began at a leading "think tank" that pioneered the original open innovation approach, where he guided initiatives for giants like Exxon and Procter & Gamble. It was there he worked alongside the creators of now-foundational concepts in portfolio management, staged development, and intrapreneurship.

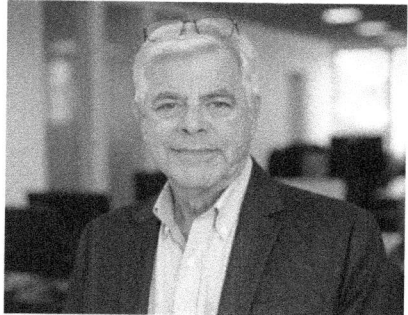

As a former president of the Product Development and Management Association (PDMA), Paul was instrumental in shaping the profession's standards, co-creating its first certification test, and establishing foundational research on measuring the success of new products. Ironically, it is these standards that this book challenges.

Today, Paul channels this vast experience into The Adept Group, a firm dedicated to helping a global client list execute smart product line strategies and accelerate their "Outcome Velocity". His expertise, featured in publications like the *Journal of Product Innovation Management* and the *PDMA Handbook of Product*

Development, underpins the Masterclasses he teaches on product line strategy and responsive roadmapping.

Following his acclaimed 2018 book, *The Profound Impact of Product Line Strategy*, this book distills his life's work into a clear, actionable guide. Paul's advice is informed not only by his MBA from Babson College and studies at Stevens Institute of Technology and MIT, but also by his hands-on work with modern AI, automation, and data transformation technologies.

In 2024, Paul solidified three critical theories to drive change in the product line: Product Line Congruency Theory, Archetype Disruptor Theory, and Product Line Change Theory. Separately and together, these theories mark a significant turning point in the management of product innovation.

www.ingramcontent.com/pod-product-compliance
Lightning Source LLC
Chambersburg PA
CBHW031837200326
41597CB00012B/178